SpringerBriefs in Molecular Science

Green Chemistry for Sustainability

Series editor

Sanjay K. Sharma, Jaipur, India

W0235042

More information about this series at http://www.springer.com/series/10045

Anjali Patel · Soyeb Pathan

Polyoxomolybdates as Green Catalysts for Aerobic Oxidation

 Springer

Anjali Patel
Polyoxometalates and Catalysis Laboratory,
  Department of Chemistry
The Maharaja Sayajirao University
  of Baroda
Vadodara, Gujarat
India

Soyeb Pathan
Polyoxometalates and Catalysis Laboratory,
  Department of Chemistry
The Maharaja Sayajirao University
  of Baroda
Vadodara, Gujarat
India

ISSN 2212-9898
ISBN 978-3-319-12987-7   ISBN 978-3-319-12988-4   (eBook)
DOI 10.1007/978-3-319-12988-4

Library of Congress Control Number: 2014954593

Springer Cham Heidelberg New York Dordrecht London

Printed on acid-free paper

Springer is part of Springer Science+Business Media (www.springer.com)

Preface

In one of the largest generic areas of chemistry (especially oxidations), there are countless processes operated by almost every type of chemical manufacturing company, producing products of incalculable value, yet producing almost immeasurable volumes of waste. Hence from the viewpoint of green chemistry, the search for efficient and heterogeneous catalytic systems exploiting clean oxidants such as molecular oxygen is highly desirable.

The present book emphasizes the different types of green heterogeneous catalysts based on phosphomolybdates as well as their use as catalysts for solvent-free oxidation of alcohols using molecular oxygen as an oxidant.

This book will be of interest to all chemists, material scientists as well as students having basic knowledge of inorganic chemistry and catalysis. It will be of enormous value to postgraduates and researchers working in the field of green chemistry.

Anjali Patel

Acknowledgments

Professor A. Patel is thankful to the Department of Science and Technology (Project No. SR/S5/GC-01/2009), New Delhi for financing a part of this work. Dr. S. Pathan is thankful to University Grants Commission (UGC-MANF & UGC-RFSMS), New Delhi for providing financial assistance. The authors are also thankful to Mr. Nilesh Narkhede (Senior Research Fellow, Council of Scientific and Industrial Research, New Delhi) and Ms. Sukriti Singh (Senior Research Fellow, Department of Science and Technology, New Delhi) for carrying out the work presented in Sect. 2.2, as well as their help in various stages in preparing the manuscript.

Contents

Abbreviations

Al_2O_3	Neutral alumina
CTAB	Cetyl trimethyl ammonium bromide
Hβ	Acidic form of zeolite β
LPOMs	Lacunary Polyoxometalates
Naβ	Sodium form of zeolite β
PMo_{11}	Undecamolybdophosphate
PMo_{11}/Al_2O_3	PMo_{11} supported on to Al_2O_3
$PMo_{11}/Hβ$	PMo_{11} anchored to Hβ
$PMo_{11}/MCM-41$	PMo_{11} anchored to MCM-41
PMo_{11}/ZrO_2	PMo_{11} supported on to ZrO_2
$PMo_{11}Co$	Mono Co(II)- substituted Phosphomolybdate
$PMo_{11}Mn$	Mono Mn(II)- substituted Phosphomolybdate
$PMo_{11}Ni$	Mono Ni(II)- substituted Phosphomolybdate
PMo_{12}	Dodecamolybdophosphate
POMs	Polyoxometalates
PTC	Phase Transfer Catalyst
TBHP	*tert*-Butyl hydrogen peroxide
TEOS	Tetra ethyl ortho silicate
TMSPMo	Transition Metal substituted Phosphomolybdates
TMSPOMs	Transition Metal substituted Polyoxometalates
TON	Turn Over Number
ZrO_2	Hydrous zirconia

Chapter 1
Introduction

Chemical industries have contributed to worldwide economic development over the past century and chemical products make an enormous contribution to the quality of our lives. At the same time, the manufacturing processes of chemicals have also led to vast amounts of wastes. Hence, today the reduction/elimination of these wastes is a central issue. Towards the same, awareness within the fine chemicals and pharmaceutical industry to improve the environmental and production costs of synthesis over the release of waste products and toxins into the environment has been increased tremendously. A sustainable society is one that meets the needs of the current generation without sacrificing the ability to meet the needs of future generations. Sustainable development is a strategic goal. It can be reached using various approaches, and this is where green chemistry comes in.

A reasonable working definition of green chemistry can be formulated as follows [1]: Green chemistry efficiently utilizes (preferably renewable) raw materials, eliminates waste and avoids the use of toxic and/or hazardous reagents and solvents in the manufacture and application of chemical products. Green chemistry deals with designing chemical products and processes that generate and use fewer (or preferably no) hazardous substances. By applying the principles of green chemistry, companies embrace cleaner and more efficient technologies, with an a priori commitment to a cleaner and healthier environment. Green chemistry eliminates waste at source, i.e. it is primary pollution prevention rather than waste remediation i.e. Prevention is better than cure.

Oxidation reactions are among the most useful and used reactions in industrial processes. In particular, the oxidation to the corresponding carbonyl compounds is of fundamental importance in organic synthesis, due to the wide ranging utility of these products as important precursors and intermediates for many drugs, vitamins and fragrances.

Because of the unreactivity of present catalysts with molecular oxygen, traditional procedures for oxidation have employed more reactive forms of oxygen i.e. toxic and hazardous oxidants based on chromates, hypochlorites, and permanganates in stoichiometric amounts or a large excess [2]. However, these procedures have some drawbacks, such as the use of relatively expensive oxidizing agents, lack of selectivity, metal waste generation and the use of non-green halogenated solvents, which are economically and environmentally undesirable. Reports are

© The Author(s) 2015
A. Patel and S. Pathan, *Polyoxomolybdates as Green Catalysts for Aerobic Oxidation*,
SpringerBriefs in Green Chemistry for Sustainability,
DOI 10.1007/978-3-319-12988-4_1

available on transition metal complexes catalyzed oxidation reactions [3–8]. At the same time the recycling of these complexes is still a challenge due to the difficult separation and rapid activity loss. Hence, from the view point of green chemistry, the search for efficient and heterogeneous catalytic systems exploiting clean oxidants such as molecular oxygen is highly desirable.

Therefore, developing selective, efficient and recyclable heterogeneous catalysts for green oxidation of organic compounds, that can use air or pure dioxygen (O_2) as oxidants under solvent free condition, is of vital importance for both economic and environmental reasons. In the recent years, air or dioxygen has emerged as the oxidant of choice for many transformations because it is environmentally benign, having high atom efficiency, it can be handled and stored safely, and it produces water, only by-product. Thus, due to the obvious advantages of using air or dioxygen over other oxidants (aqueous H_2O_2, TBHP, PhOI etc.) as the crucial and stoichiometric oxidant, considerable effort has been invested in the last few years to develop novel catalysts for the aerobic oxidation of organic substrates to their corresponding oxygenated products. In this context, Polyoxometalates (POMs), have been gaining importance due to their redox properties [9–14].

POMs are a distinctive class with unique properties of topology, size, electronic versatility as well as structural diversity. Due to the combination of their added value properties such as redox properties, large sizes, high negative charge, nucleophilicity they play a great role in various fields such as medicine, material science, photochromism, electrochemistry, magnetism as well as catalysis. POMs are a rich class of inorganic metal-oxide cluster compounds with transition metals in their highest oxidation state and have general formula $[X_xM_mO_y]^{n-}$, in which X is the hetero atom, usually a main group element (e.g. P, Si, Ge, As), and M is the addenda atom, being a d-block element in high oxidation state, usually V^V, Mo^{VI} or W^{VI} [15, 16]. These compounds are always negatively charged, although the negative density is widely variable depending on the elemental composition and the molecular structure.

The POMs have been known since the first report by Berzelius. He described the yellow precipitate (ammonium 12-molybdophosphate) that is produced when ammonium molybdate is added in excess to phosphoric acid [17]. After the discovery of this first POM, the field of POM chemistry progressed significantly [18–27] and various types of structures were discovered. After that, an extensive literature on their synthesis and structure has been accumulated and summarized in the form of reviews as well as books. Polyhedral representations of various types of POMs are presented in Fig. 1.1.

Among different POMs, Keggin type POMs are investigated extensively because of their easy synthesis as well as stability [16]. The ideal Keggin structure, $[XM_{12}O_{40}]^{3-}$ of α-type has Td symmetry and consists of a central XO_4 tetrahedron (X = heteroatom or central atom) surrounded by twelve MO_6 octahedra (M = addenda atom). The twelve MO_6 octahedra comprise four groups of three edge-shared octahedra, the M_3O_{13} triplet [22, 23], which have a common oxygen vertex connected to the central heteroatom. The oxygen atoms in this structure fall into four classes of symmetry-equivalent oxygens: $X-O_a-(M)_3$, $M-O_b-M$, connecting two

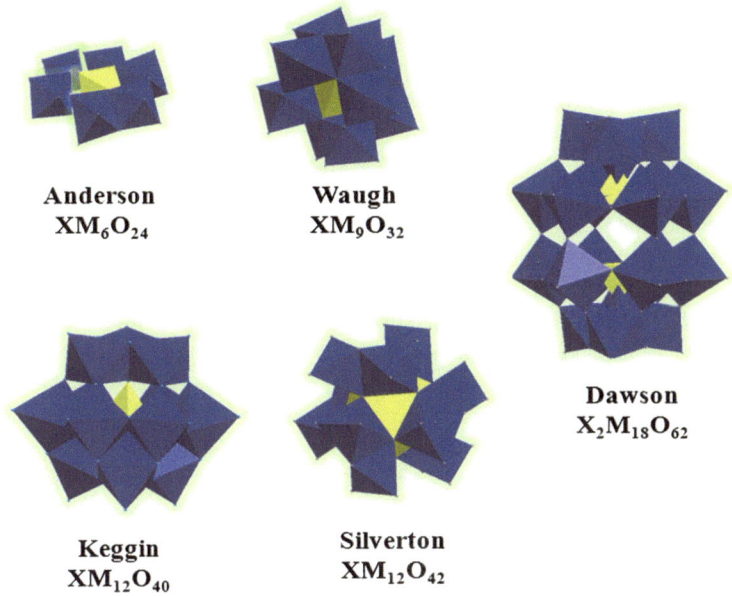

Fig. 1.1 POMs family (polyhedral representation)

M_3O_{13} units by corner sharing; $M-O_c-M$, connecting two M_3O_{13} units by edge sharing; and O_d-M, where M is the addenda atom and X the heteroatom. The schematic representation of Keggin type POM is shown in Fig. 1.2.

Depending on the solvent, the acidity of the solution and the charge of the polyanion, the reductions involve either single electron or multi electron steps often accompanied by protonation. The oxidation potential is strongly dependent on the addenda atom and is not much influenced by central hetero atom. The oxidation potentials of polyanions containing Mo and V are high as these ions are easily reduced. It has been reported that oxidative ability decreases in the order $V^- > Mo^- > W^-$ containing heteropolyanion. The basic importance of using POMs in various homogeneous as well as heterogeneous oxidation catalysis is due to their inherent stability towards strong oxidants as well as their ability to retain their

Fig. 1.2 Keggin type POM; $[XM_{12}O_{40}]$ anion

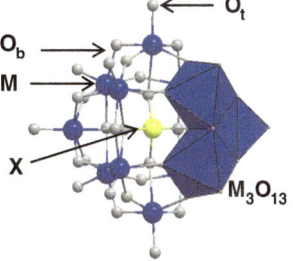

structures at high temperatures. They are widely used as a model system for fundamental research providing unique opportunities for mechanistic studies on the molecular level. At the same time they become increasingly important for applied catalysis. They provide good basis for the molecular design of mixed oxide catalyst and they have high capability in practical uses.

The first attempts to use POMs as catalysts were tracked way back in the 19th century. Systematic investigation of catalysis by POMs began in the early 1970s. Some of the major achievements of POM based compounds in the field of catalysis have been reviewed by number of groups [28–39]. Numbers of patents describing use of POMs based compounds in catalysis are also available [40–54]. The world of catalysis by POMs is largely expanded and it would be difficult to mention all references. Hence, in the present report, we would like to restrict ourselves to Keggin type polyoxomolybdates, especially for oxidation catalysis.

It is well known that the redox properties of POMs can be tuned at molecular level which can lead to development of a new class of materials with unique structural as well as electronic properties. One of the most significant properties of modified precursors is their ability to accept and release specific numbers of electrons reversibly, under marginal structural rearrangement [55–57]. Thus, the modification of parent POMs are likely to help in development of new generation catalysts with enhanced redox properties as well as stability.

The modification of properties can be basically done by tuning the structural properties at the atomic or molecular level in two ways (i) By creating defect (lacuna) in parent POM structures (i.e. Lacunary Polyoxometalates) and (ii) Incorporation of transition metal ions into the defect structures (i.e. Transition Metal Substituted Polyoxometalates).

1.1 Lacunary Polyoxometalates (LPOMs)

LPOMs are a sub class of POMs with a set of unique properties such as mult-identicity, rigidity, thermal and oxidative stability [18, 58, 59]. They represent an important class of compounds due to their unique structural as well as chemical properties. Controlled treatment of heteropoly/polyoxo species with base can produce "lacunary" heteropoly/polyoxo species wherein one or more addenda atoms have been eliminated from the structure along with the oxygens [60]. Due to the structural diversity as well as the unique electronic properties, LPOMs are of potential importance.

Removal of one or two MO units from the fully occupied POMs, $[XM^{VI}_{12}O_{40}]^{n-}$, gives rise to mono- or di-lacunary POMs, $[XM^{VI}_{11}O_{39}]^{(n+4)-}$ and $[XM^{VI}_{10}O_{36}]^{(n+5)-}$.

The classical examples of geometries observed in LPOMs are shown in Fig. 1.3.

Formation of mono, di or tri lacunary species is mainly pH dependent, each possessing its own reactivity and stability trend. Hence, synthetically, special attention is paid to fine changes in reaction conditions such as pH, temperature,

Fig. 1.3 Different types of
LPOMs structures derived
from parent Keggin unit
a $[XM_{11}O_{39}]^{n-}$, **b** $[XM_{10}O_{36}]^{n-}$, **c** A-$[XM_9O_{34}]^{n-}$
d B-$[XM_9O_{34}]^{n-}$

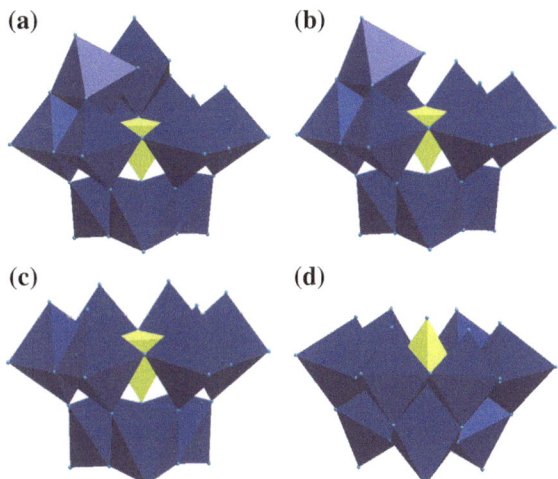

buffer capacity, ionic strength, and cation size: all having the potential to exert a considerable effect on the polyanion equilibria and formation of products [61, 62].

Among the large variety of LPOMs, the mono lacunary Keggin type POMs form the most versatile class of LPOMs. As mentioned earlier, the removal of one MO at suitable pH from parent POMs leads to the formation of mono LPOMs (Fig. 1.4).

A list of mono LPOMs are represented in Table 1.1.

A literature survey shows that system based on lacunary polyoxotungstates, $[XW_{11}O_{39}]^{n-}$ (X = Si, P) are well documented [63–74]. At the same time, reports on the analogous compounds, based on mono lacunary polyoxomolybdates are scarce.

In 1991, isolation and purification of lacunary polyoxomolybdates $[PMo_{11}O_{39}]^{7-}$ as tetrabutyl ammonium salt was successfully achieved by Combs-Walker and Hill [75]. The synthesized material was well characterized by FT-IR, UV-visible spectroscopy, ^{31}P NMR and cyclic voltammetry studies. In addition, a literature survey also shows that no study has been carried out in last two decades on the $[PMo_{11}O_{39}]^{7-}$ due to the difficulty in isolation, and poor thermal as well as kinetic stability of the same.

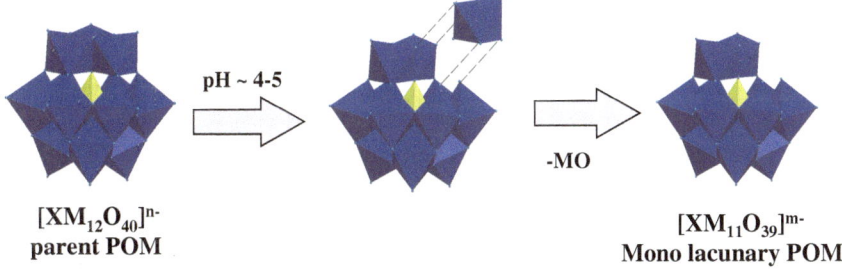

$[XM_{12}O_{40}]^{n-}$
parent POM

pH ~ 4-5

-MO

$[XM_{11}O_{39}]^{m-}$
Mono lacunary POM

Fig. 1.4   Formation of mono LPOMs

Table 1.1 Different LPOMs with variable heteroatoms [58]

	X	Isomer
$[XW_{11}O_{39}]^{n-}$	P, As	α
	Si, Ge	$\alpha, \beta_1, \beta_2, \beta_3$
	B	–
	Al, Ga, Fe(III), Co(III)	α
	Co(II), Zn	α^a
	Sb(III), Bi(III)	–
$[XMo_{11}O_{39}]^{n-}$	P, As, Si, Ge	α

[a] Free ligand structure not known

The present book focuses on the isolation as well as stabilization of $[PMo_{11}O_{39}]^{7-}$ specie. The stability of $[PMo_{11}O_{39}]^{7-}$ was enhanced by making it heterogeneous after supporting on to suitable supports (Chap. 2). Supports not only stabilize $[PMo_{11}O_{39}]^{7-}$ but also modify the properties of active species by providing an opportunity to disperse over a large surface area. The nature of the support plays an important role, especially for a successful catalytic reaction. Hence Metal oxides (Hydrous zirconia; ZrO_2 and neutral alumina; Al_2O_3) and porous materials (MCM-41 and zeolite-Hβ) were used as supports. Use of supported $[PMo_{11}O_{39}]^{7-}$ as catalysts for selective aerobic oxidation of alcohols was also discussed. Scope and limitation of synthesized materials as heterogeneous catalysts was also shown for oxidation of different alcohols. Catalytic activity was also correlated with the nature of support.

1.2 Transition Metal Substituted Polyoxometalates (TMSPOMs)

The second way of modification of POMs is via incorporation of transition metal ions into the defect structures, forming a new class of compounds known as TMSPOMs. TMSPOMs have received increasing attention because substitution of a transition metal into the POM has been explored as a route to increase the range of application of these compounds [76–79]. This mostly due to the fact that TMSP-OMs can be rationally modified at the molecular level including size, shape, charge density, acidity, redox states, stability and solubility resulting in their outstanding chemical properties. It affords convenient platforms for the stabilization of unusually high oxidation state metal-oxo species. In addition, TMSPOMs have number of advantages over organometallic complexes such as (i) their solubility is tunable by changing counter cations (ii) their redox properties are adjustable by changing the central (hetero) atom and the incorporated transition metal (iii) they are robust under oxidation conditions, under which most organic ligands decompose [80].

This unique class of metal oxygen clusters, are outstanding inorganic building blocks due to their undisputed structural beauty and controllable sizes, shapes and high negative charges [59, 81–87]. They exhibit an enormous variety of structures, which leads to interesting and unexpected properties that give rise to many applications in magnetism, medicine and catalysis [32, 88].

Most of the known TMSPOMs are mono-, di-, or trisubstituted derivatives of the Keggin type POM, e.g. $[XM_{11}M'O_{39}]^{n-}$, $[XM_{10}M_2'O_{38}]^{m-}$ and $[XM_9M_3'O_{37}]^{p-}$. Among these, mono-substituted Keggin derivatives $[XM_{11}M'O_{39}]^{(n-m)-}$ (X = P, Si, B; M = W, Mo; M' = transition metal), are explored extensively. Mono TMSPOMs recognized as inorganic analogs of metalloporphyrin complexes [89]. They have distinct advantages over the metalloporphyrin and organometallic complexes such as hydrolytically stable and thermally robust. The non-oxidizable M-oxo framework of polyanions acts as an inert, multi-dentate ligand which can accommodate a multitude of transition metal centers [89–92].

The reaction of the mono lacunary $\{XM_{11}O_{39}\}^{m-}$ (where X = P, B, Si or Ge; M = W^{VI}, Mo^{VI}, $V^{V,VI}$, etc.) anion with transition-metal cations M' in aqueous solution leads to the formation of mono TMSPOMs $\{XM_{11}O_{39}M'(L)\}$, (where M' = d-electron transition metal and L = monodentate ligand, generally H_2O) (Fig. 1.5).

The environment around the substituent M' in the metal substituted polyanions $[XM_{11}O_{39}M'(H_2O)]^{n-}$ is considered to be near octahedral. In particular, the complexes of the type $[XM_{11}O_{39}M'(H_2O)]^{n-}$ show many analogies to metalloporphyrins [57, 89, 93]. TMSPOMs based materials have played an important role as catalysts in the field of oxidation catalysis due to their tendency to exhibit fast reversible multi-electron redox transformations under rather mild conditions and their inherent stability towards strong oxidants. Hence they are widely used as a model system for fundamental research providing unique opportunities for mechanistic studies on the molecular level. At the same time they become increasingly important for applied catalysis.

Majority of applied work was carried out in field of catalysis, especially oxidation of various organic substrates. Because the catalytically active site is at the substituted transition metal centre and POMs functions as a ligand with a strong capacity for accepting electrons. The substituting metal center is thus pentacoordinated by the

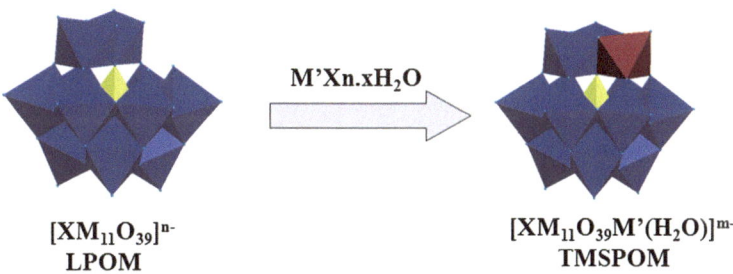

$M'Xn.xH_2O$

$[XM_{11}O_{39}]^{n-}$
LPOM

$[XM_{11}O_{39}M'(H_2O)]^{m-}$
TMSPOM

Fig. 1.5 Formation of TMSPOMs from LPOMs

"parent" POMs. The octahedral coordination sphere is completed by an additional sixth labile ligand, L (usually L = H_2O). This lability of the sixth ligand allows the interaction of the substituting transition metal atom reacting with substrate and/or oxidant. In analogy with organometallic chemistry the "pentadentate" POMs acts as an inorganic ligand. This analogy led to transition metal-substituted polyoxometalates being termed "inorganic metalloporphyrins" and has distinct advantages over organometallic species, e.g. they are rigid, hydrolytically stable and thermally robust. Further, the "active sites" of their transition metals and countercations can undergo extensive synthetic modifications.

Numbers of groups have given a significant contribution in this field of oxidation catalysis. They have established unique and efficient catalytic systems based on $[XW_{11}O_{39}M(H_2O)]$ (X = P, Si; M = transition metal) as catalysts with various oxidants such as O_2, H_2O_2, PhOI, TBHP, $NaIO_4$ [76, 94–108]. In the present book we have restricted ourselves to mono transition metal substituted phosphomolybdates, (TMSPMo), mainly because of two reasons (i) all available number of references on TMSPW would result in increasing the bulk of the report (ii) TMSPMo have generated substantial interest as oxidation catalysts.

Among TMSPMo, numbers of reports are available on $[PV(H_2O)Mo_{11}O_{39}]^{4-}$ based catalysts for oxidation/hydroxylation of benzene with H_2O_2 by Gopinathan et al. [109], Hu et al. [110], Nomiya et al. [111], Mizuno et al. [112], You et al. [113], Wang et al. [114] and Zhang et al. [115]. Similarly, oxidative dehydrogenation of isobutyric acid over $[PV(H_2O)Mo_{11}O_{39}]^{4-}$ based catalysts has also been reported by Hervé et al. [116, 117], Kaiji et al. [118], Akimoto et al. [119] and Misono et al. [120]. Oxidative dehydrogenation of propane catalyzed by Cs salt of $[PVMo_{11}O_{40}]$ has also been reported by Vedrine and co-worker [121, 122] and Wan et al. [123]. Mizuno et al. [124] reported aerobic oxidation of adamantane over $[PV(H_2O)Mo_{11}O_{39}]$ based catalyst at 356 K. Vapour phase oxidation of n-Pentane over $H_4[PVMo_{11}O_{40}]$ has been reported by Centi et al. [125]. $H_4[PVMo_{11}O_{40}]$ catalyzed oxygenation of Methane with H_2O_2 in $(CF_3CO)_2O$ solvent at 353 K has been reported by Mizuno and group [126, 127]. Oxidation of isobutane on $(NH_3)_3H$ $[PVMo_{11}O_{40}]$ [128], $Cs_{2.5}Ni_{0.08}H_{1.34}[PVMo_{11}O40]$ [129, 130], $H_4[PVMo_{11}O_{40}]$ [131], $(NH_4)_3HPMo_{11}VO_{40}$ [132] has also been reported.

Kholdeeva and co-worker [133] reported oxidation of 2,3,6-trimethylphenol catalyzed by $[PVMo_{11}O_{40}]^{4-}$ based catalyst in presence of organic: H_2O solvent system. Oxidation of phenol with air using $[(C_nH_{2n+1})N(CH_3)_3]_{3+x}[PVxMo_{12-x}O_{40}]$ (n = 8–18; x = 1, 2, 3) at room temperature has been reported by Huo et al. [134].

Series of cesium salt of $[PVMo_{11}O_{40}]$ have been used as catalysts for gas phase oxidation of propylene by Dimitratos et al. [135]. Lee et al. [136], Hibst et al. [137] and Zhang et al. [138] explored use of Cs salt of $[PVMo_{11}O_{40}]$ for Methacrolein oxidation. Zhizhina et al. [139] reported aerobic oxidation of propene and butane-1 using $Pd+H_4[PVMo_{11}O_{40}]$ as a catalysts at 333 K in homogeneous medium. Reports on the oxidation of norbornene with aqueous H_2O_2 as an oxidant in different solvents, [140] and epoxidation of different alkenes with TBHP/H_2O_2 in CH_3CN and CH_2Cl_2 [141] over $[PVMo_{11}O_{40}]^{4-}$ are also available. Vapour-phase oxidation of propene over $H_{3+x}PMo_{12-x}V_xO_{40}$ based catalysts has also been reported [142]. Lee et al. [12] established use of $Cs_nH_{4-n}[PVMo_{11}O_{40}]$ as active and selective catalysts for vapour phase oxidation of ethanol. Peng et al. [10] studied the oxidation of liquid phase benzyl alcohol over a series of Cs salts of $[PVMo_{11}O_{40}]^{4-}$ using H_2O_2 as oxidant. Number of reports have been reported on oxidation reactions using $[PVMo_{11}O_{40}]$ as catalysts by Sai Prasad and group [143–148].

Thus, a literature survey shows that most of reports described catalytic of $[PVMo_{11}O_{40}]^{4-}$ based materials. At the same time report on mono transition metal substituted polyoxomolybdates, $[XMo_{11}O_{39}M(H_2O)]$ (X = P, Si; M = transition metal except vanadium) are very less.

Combs-Walker and Hill [75] reported two steps synthesis of $TBA_4H[PMo_{11}TM(L)O_{39}]$ (TM = Co, Mn, Cu and Zn; TBA = tetra butyl ammonium). Using same method, Neumann et al. [149] reported synthesis of TBA salt of $[PMo_{11}Ru(L)O_{39}]$. They reported use of synthesized material as bifunctional catalysts for the aerobic oxidation of cumene to hydroperoxo/peroxo intermediate followed by oxygen transfer to an alkene to yield epoxide in acetonitrile. Oxidative dehydrogenation of 2-propanol over $Cs_{2.5}H_{1.2}PMo_{11}Fe(H_2O)O_{39} \cdot 6H_2O$ has also been reported by Mizuno et al. [150].

Rabia et al. [151] reported synthesis of NH_4-salt of $[PMo_{11}MO_{40}]$ (M = Co, Ni, Fe) and their use a catalyst for oxidation of propane with molecular oxygen at temperature 380–420 °C. Hue and Burns reported synthesized and characterized a series of $Na_2[(C_4H_9)_4N]_4[PZ(II)(Br)Mo_{11}O_{39}]$, where Z = Mn(II), Co(II), Ni(II), Cu (II) and Zn(II)). They explored the use of synthesized materials as catalysts for homogeneous oxidation of isobutyraldehyde using H_2O_2 as an oxidant in acetonitrile at 50 °C [152].

Vázquez and group [153] established use of $H_6PMo_{11}BO_{40}$, $H_6PMo_{11}BiO_{40}$, $H_6PMo_{11}LaO_{40}$, and $H_6PMo_{11}YO_{40}$ as catalysts in the green and selective oxidation of diphenyl sulphide with H_2O_2.

Use of TBA-salt of $[PRu(H_2O)Mo_{11}O_{39}]^{4-}$ as catalysts for the epoxidation of alkenes with molecular oxygen in acetonitrile under mild condition has been reported by Naumann and co-worker [154]. Oxidative dehydrogenation of alcohols under aerobic as well as anaerobic conditions over $[PSb^V(O)Mo_{11}O_{39}]^{4-}$ and $[PSb^V(Br)Mo_{11}O_{39}]^{3-}$ in benzonitrile at 135 °C has also been investigated by the same group [155]. Knapp et al. [156] reported vapour phase selective oxidation of isobutene over $Cs_3H_1PMo_{11}FeO_{39}$. Tundo et al. [157] reported multiphase oxidation of alcohols and sulfides with hydrogen peroxide catalyzed by $H_6PMo_{11}AlO_{40}$ at 70 °C. Use of TBA salt of $PMo_{11}MO_{40}$ (M = Co, Mn) as catalysts for oxidation of phenol in acetonitrile has been explored by Karcz et al. [158]. Recently, Cavaleiro et al. [159] reported homogeneous catalytic oxidation of olefins with hydrogen peroxide catalyzed by $TBA_4H[PMo_{11}Mn(H_2O)O_{39}] \cdot 2H_2O$ in acetonitrile.

From literature survey it can be seen than, in most of applied protocols catalyzed by TMSPMo, oxidation requires harsh conditions of temperature, use of relatively expensive metals (V, Ru, Bi, La, Y) as well as use of organic solvent.

Considering these aspects, it was thought of interest to summarize and describe a new easy catalytic pathway with TMSPMo, and to discuss their use as sustainable catalysts for oxidation reactions. As Mn, Co, Ni and Cu are most important from the view point of their redox properties, they were selected for the present work.

Chapter 3 describes synthesis and characterization transition metal (Mn, Co, Ni Cu)-substituted phosphomolybdates and their use as heterogeneous catalysts for solvent free aerobic oxidation of alcohols under mild reaction conditions. Scope and limitation of synthesized materials was also discussed for oxidation of different alcohols. Based on results, possible reaction mechanism was also explained.

It is worth to notice that all the reported catalysts in the book are reusable and are promising environmentally benign catalysts.

References

1. Sheldon RA (2000) C R Acad Sci Paris IIc Chimie/Chemistry 3:541–551
2. Sheldon RA (1991) Stud Surf Sci Catal 59:33–54
3. Piera J, Bäckvall J-E (2008) Angew Chem Int Ed 47:3506–3523
4. Parmeggiani C, Cardona F (2012) Green Chem 14:547–564
5. Sheldon RA, Arends IWCE, Dijksman A (2000) Catal Today 57:157–166
6. Komiya N, Nakae T, Sato H, Naota T (2006) Chem Commun 46:4829–4831
7. Wang Q, Zhang Y, Zheng G, Tian Z, Yang G (2011) Catal Commun 14:92–95
8. Zhao G, Hu H, Deng M, Lu Y (2011) Chem Commun 47(34):9642–9644
9. Ding Y, Ma B, Gao Q, Li G, Yan L, Suo J (2005) J Mol Catal A Chem 230:121–128
10. Peng G, Wang Y, Hu C, Wang E, Feng S, Zhou Y, Ding H, Liu Y (2001) Appl Catal A Gen 218:91–99
11. Li M, Shen J, Ge X, Chen X (2001) Appl Catal A Gen 206:161–169
12. Yang JH, Lee DW, Lee JH, Hyun JC, Lee KY (2000) Appl Catal A Gen 194–195:123–127
13. Li W, Lee K, Oshihara K, Ueda W (1999) Appl Catal A Gen 182:357–363
14. Mizuno N, Tateishi M, Iwamoto M (1996) J Catal 163:87–94
15. Mizuno N, Hikichi S, Yamaguchi K, Uchida S, Nakagawa Y, Uehara K, Kamata K (2006) Catal Today 117:32–36
16. Pope MT, Muller A (eds) (2001) Polyoxometalate chemistry: from topology via self-assembly to applications. Kluwer, Dordrecht
17. Berzelius JJ (1826) Pogg Ann Phys Chem 6:369–392
18. Pope MT, Muller A (2003) Polyoxometalate molecular science. Kluwer Academic Publishers, Dordrecht
19. Svanberg L, Struve H (1848) J Prakt Chem 44:257–291
20. Miolati A, Pizzighelli R (1908) J Prakt Chem 77:417–456
21. Rosenheim A, Jaenicke H (1917) Z Anorg Allg Chem 100:304–354
22. Pauling LC (1929) J Am Chem Soc 51:2868–2880
23. Keggin JF (1933) Nature 131:908–909
24. Keggin JF (1934) Proc Roy Soc A 144:75–100
25. Signer R, Gross H (1934) Helv Chim Acta 17:1076–1080
26. Bradley AJ, Illingworth JW (1936) Proc R Soc A157:113–131
27. Brown GM, Noe-Spirlet M-R, Busing WR, Levy HA (1977) Acta Crystallogr B 33:1038–1046
28. Corma A (1995) Chem Rev 95(3):559–614
29. Okuhara T, Mizuno N, Misono M (1996) Adv Catal 41:113–252
30. Kozhevnikov IV (1998) Chem Rev 98(1):171–198

31. Mizuno N, Misono M (1998) Chem Rev 98(1):199–218
32. Kozhevnikov IV (2002) Catalysts for fine chemical synthesis: catalysis by polyoxometalates, vol 2. Wiley, New York
33. Mallat T, Baiker A (2004) Chem Rev 104:3037–3058
34. Sheldon RA, Arends I, Hanefeld U (2007) Green chemistry and catalysis. Wiley-VCH Weinheim, Germany
35. Mizuno N, Yamaguchi K, Kamata K, Nakagawa Y (2008) In: Oyama ST (ed) Mechanisms in homogeneous and heterogeneous epoxidation catalysis. Elsevier Publications, Amsterdam, pp 155–176
36. Mizuno N, Kamata K, Uchida S, Yamaguchi K (2009) In: Mizuno N (ed) Modern heterogeneous oxidation catalysis. Wiley-VCH Weinheim, Germany, pp 185–216
37. Ren Y, Yue B, Gu M, He H (2010) Materials 3:764–785
38. Clerici MG, Kholdeeva OA (eds) (2013) Liquid phase oxidation via heterogeneous catalysis: organic synthesis and industrial applications. Wiley Inc., Hoboken, New Jersey
39. Patel A (ed) (2013) Environmentally benign catalysts for cleaner organic reactions. Springer Inc., New York
40. Lyons JE, Ellis PE Jr, Langdale WA, Myers HK Jr (1990) US Patent 4916101 A, 10 Apr 1990
41. Knifton JF (1994) US Patent 5300703, 5 Apr 1994
42. Kresge CT, Marler DO, Rav GS, Rose BH (1994) US Patent 5324881, 28 June 1994
43. Soled SL, McVicker GB, Niseo S, Gates WE (1995) US 5391532, 21 Feb 1995
44. Angstadt HP, Hollstein EJ, Hsu C-Y (1996) US 5493067, 20 Feb 1996
45. Rhubright DC, Burrington JD, Zhu PY (1998) US 5817831, 6 Oct 1998
46. Davis ME, Dillon CJ, Holles JH, Labinger JA, Brait A (2005) US 6914029 B2, 5 July 2005
47. Liu H, Iglesia E (2005) US 6956134 B2, 18 Oct 2005
48. Bailey C, Gracey B (2008) US 2008/004466 A1, 3 Jan 2008
49. Patel A, Bhatt N (2009) US 7692047 B2, 6 Apr 2010
50. Gracey BP, Haining GJ, Partington SR (2010) US 2010/0292520A1, 18 Nov 2010
51. Okun N, Hill CL (2010) US 7655594 B2, 2 Feb 2010
52. Voitl T, Von Rohr PR (2011) US 7906687, 15 Mar 2011
53. Bhotla VRNG, Halligudi SB, Kishan G, Mallika SP, Veldurthy B (2011) US 7868190 B2, 11 Jan 2011
54. Jaensch H, Dakka JM, Benitez FM, Kortz U, Richards RM (2011) US 7906686 B2, 15 Mar 2011
55. Weinstock IA (1998) Chem Rev 98:113–170
56. Sadakane M, Steckhan E (1998) Chem Rev 98:219–237
57. Thouvenot R, Proust A, Gouzerh P (2008) Chem Comm 16:1837–1852
58. Pope MT (1983) Heteropoly and isopoly oxometalates. Springer-Verlag, Berlin
59. Pope MT, Müller A (1991) Polyoxometalate chemistry: an old field with new dimensions in several disciplines. Angew Chem Int Ed Engl 30:34–48
60. Baker LCW, Glick D (1998) Chem Rev 98:3–50
61. Nsouli NH, Ismail AH, Helgadottir IS, Dickman MH, Clemente-Juan JM, Kortz U (2009) Inorg Chem 48:5884–5890
62. Mitchell SG, Miras HN, Long D, Cronin L (2010) Inorg Chim Acta 363:4240–4246
63. Matsumoto K, Sasaki Y (1976) Bull Chem Soc Jpn 49:156–158
64. Brevard C, Schimpf R, Tourne G, Tourne C (1983) J Am Chem Soc 105:7059–7063
65. Honma N, Kusaka K, Ozeki T (2002) Chem Comm 2896–2897
66. Laurencin D, Proust A, Gerard H (2008) Inorg Chem 47:7888–7893
67. Shringarpure P, Patel A (2008) Dalton Trans 3953–3955
68. Shringarpure P, Patel A (2010) Dalton Trans 39:2615–2621
69. Pathan S, Patel A (2010) J Coord Chem 63:4041–4049
70. Shringarpure P, Patel A (2011) Chem Eng J 173:612–619
71. Shringarpure P, Patel A (2011) Ind Eng Chem Res 50:9069–9076

72. Patel A, Narkhede N (2013) Catal Sci Technol 3:3317–3325
73. Narkhede N, Patel A, Singh S (2014) Dalton Trans 43:2512–2520
74. Patel A, Singh S (2014) Microporous Mesoporous Mater 195:240–249
75. Combs-Walker LA, Hill CL (1991) Inorg Chem 30:4016–4026
76. Hill CL, Brown RB (1986) J Am Chem Soc 108:536–538
77. Toth JE, Melton JD, Cabelli D, Bielski BHJ, Anson FC (1990) Inorg Chem 29:1952–1957
78. Rong C, Anson FC (1994) Inorg Chem 33:1064–1070
79. Muller A, Dloczik L, Dittman E, Pope MT (1997) Inorg Chim Acta 257:231–239
80. Sadakane M, Tsukuma D, Dickmann M, Bassil B, Kortz U, Higashijima M, Ueda W (2006) Dalton Trans 35:4271–4276
81. Hill CL (1998) Chem Rev 98(1) (Issue on "polyoxometalates")
82. Hagrman PJ, Hagrman D, Zubieta J (1999) Angew Chem Int Ed 38:2638–2684
83. Burkholder E, Zubieta J (2001) Chem Commun 20:2056–2057
84. Wu CD, Lu CZ, Zhuang HH, Huang JS (2002) J Am Chem Soc 124:3836–3837
85. Fukaya K, Yamase T (2003) Angew Chem Int Ed 42:654–658
86. Tan HQ, Li YG, Zhang ZM, Qin C, Wang XL, Wang EB, Su ZM (2007) J Am Chem Soc 129:10066–10067
87. Thiel J, Ritchie C, Streb C, Long DL, Cronin L (2009) J Am Chem Soc 131:4180–4181
88. Pope MT, Müller A (eds) (1994) Polyoxometalates: from platonic solids to anti-retroviral activity. Kluwer, Dordrecht
89. Bi L-H, Reicke M, Kortz U, Keita B, Nadjo L, Clark RJ (2004) Inorg Chem 43:3915–3920
90. Hussain F, Bassil BS, Bi L-H, Reicke M, Kortz U (2004) Angew Chem Int Ed 43:3485–3488
91. Kortz U, Nellutla S, Stowe AC, Dalal NS, Rauwald U, Danquah W, Ravot D (2004) Inorg Chem 43:2308–2317
92. Kortz U, Nellutla S, Stowe AC, Dalal NS, Tol VJ, Bassil BS (2004) Inorg Chem 43:144–154
93. Katsoulis DE, Pope MT (1986) J Chem Soc Chem Comm 15:1186–1188
94. Faraj M, Hill C (1987) Chem Commun 19:1487–1489
95. Neumann R (1989) Chem Commun 18:1324–1325
96. Neumann R, Abu-Gnim C (1990) J Am Chem Soc 112:6025–6031
97. Lyons JE, Ellis PE Jr, Durante VA, Grasselli RA, Sleight AW (eds) (1991) Stud Surf Sci Catal. Elsevier Scientific, Amsterdam
98. Lyons JE, Ellis PE Jr, Myers HK Jr, Suld G, Langdale WA (1989) US 4803187, 7 Feb 1989
99. Harrup MK, Hill CL (1994) Inorg Chem 33:5448–5455
100. Mansuy D, Bartoli J, Battioni P, Lyon D, Finke R (1991) J Am Chem Soc 113:7222–7226
101. Neumann R, Simandi LI (eds) (1991) Dioxygen activation and homogeneous catalytic oxidation. Elsevier Science Publishers, Amsterdam
102. Mizuno N, Hirose T, Tateishi M, Iwamoto M (1993) Chem Lett 11:1839–1842
103. Shringarpure P, Patel A (2014) Synth React Inorg Metal-Org Nano-Met Chem (Accepted)
104. Patel A, Patel K (2014) Polyhedron 69:110–118
105. Patel K, Shringarpure P, Patel A (2012) Supramol Chem 24:149–156
106. Patel K, Shringarpure P, Patel A (2011) Trans Met Chem 36:171–177
107. Shringarpure P, Patel A (2011) J Clus Sci 22:587–601
108. Shringarpure P, Patel A (2009) Inorg Chim Acta 362:3796–3800
109. Alekar NA, Indira V, Halligudi SB, Srinivas D, Gopinathan S, Gopinathan C (2000) J Mol Catal A Chem 164:181–189
110. Zhang J, Tang Y, Li G, Hu C (2005) Appl Catal A Gen 278:251–261
111. Nomiya K, Yagishita K, Nemoto Y, Kamataki T (1997) J Mol Catal A Chem 126:43–53
112. Misono M, Mizuno N, Inumaru K, Koyano G, Lu X-H (1997) Stud Surf Sci Catal 110:35–42
113. Yang H, Wu Q, Li J, Dai W, Zhang H, Lu D, Gao S, You W (2013) Appl Catal A Gen 457:21–25
114. Zhang F, Guo M, Ge H, Wang J (2007) Chinese J Chem Eng 15:895–898
115. Leng Y, Wang J, Zhu D, Shen L, Zhao P, Zhang M (2011) Chem Eng J 173:620–626
116. Laronze N, Marchal-Roch C, Guillou N, Liu FX, Hervé G (2003) J Catal 220:172–181

117. Roch CM, Laronze N, Guillou N, Teze A, Hervé G (2000) Appl Catal A Gen 199:33–44
118. Tonghao W, Yuchun L, Hongmao Y, Guojia W, Hengbin Z, Shiying H, Yuzi J, Kaiji Z (1989) J Mol Catal 57:193–200
119. Akimoto M, Ikeda H, Okabe A, Echigoya E (1984) J Catal 89:196–208
120. Lee KY, Oishi S, Igarashi H, Misono M (1997) Catal Today 33:183–189
121. Dimitratos N, Védrine JC (2003) Appl Catal A Gen 256:251–263
122. Dimitratos N, Védrine JC (2006) J Mol Catal A Chem 255:184–192
123. Sun M, Zhang J, Cao C, Zhang Q, Wang Y, Wan H (2008) Appl Catal A Gen 349:212–221
124. Shinachi S, Matsushita M, Yamaguchi K, Mizuno N (2005) J Catal 233:81–89
125. Centi G, Nieto JL, Iapalucci C, Bruckman K, Serwicka EM (1989) Appl Catal 46:197–212
126. Seki Y, Mizuno N, Misono M (1997) Appl Catal A Gen 158:L47–L51
127. Seki Y, Min JS, Misono M, Mizuno N (2000) J Phys Chem B 104(25):5940–5944
128. Jing F, Katryniok B, Richard EB, Paul S (2013) Catal Today 203:32–39
129. Mizuno N, Han W, Kudo T, Iwamoto M (1996) Stud Surf Sci Catal 101:1001–1010
130. Mizuno N, Yahiro H (1998) J Phys Chem B 102(2):437–443
131. Paul S, Chu W, Sultan M, Bordes-Richard E (2010) Sci China Chem 53:2039–2046
132. Jing F, Katryniok B, Dumeignil F, Bordes-Richard E, Paul S (2014) J Catal 309:121–135
133. Kholdeeva OA, Golovin AV, Makisimovskaya RI, Kozhenikov IV (1992) J Mol Catal 75:235–244
134. Zhao S, Wang X, Huo M (2010) Appl Catal B: Env 97:127–134
135. Dimitratos N, Pina CD, Falletta E, Bianchi CL, Dal SV, Rossi M (2007) Catal Today 122:307–316
136. Stytsenko VD, Lee WH, Lee JW (2001) Kinet Catal 42:212–216
137. Deusser LM, Petzoldt JC, Gaube JW, Hibst H (1998) Ind Eng Chem Res 37:3230–3236
138. Zhang H, Yan R, Yang L, Diao Y, Wang L, Zhang S (2013) Ind Eng Chem Res 52:4484–4490
139. Zhizhina EG, Simovova MV, Odyakov VF, Matveev KI (2007) Appl Catal A Gen 319:91–97
140. Kala Raj NK, Ramaswamy AV, Manikandan P (2005) J Mol Catal A Chem 227:37–45
141. Benlounes O, Cheknoun S, Mansouri S, Rabia C, Hocine J (2011) J Taiwan Inst Chem Eng 42:132–137
142. Benadji S, Eloy P, Leonard A, Su B-L, Rabia C, Gaigneaux EM (2012) Micropor Mesopor Mater 154:153–163
143. Lingaiah N, Reddy KM, Babu NS, Rao KN, Suryanarayana I, Sai Prasad PS (2006) Catal Commun 7:245–250
144. Nagaraju P, Pasha N, Sai Prasad PS, Lingaiah N (2007) Green Chem 9:1126–1129
145. Reddy KM, Balaraju M, Sai Prasad PS, Suryanarayana I, Lingaiah N (2007) Catal Lett 119:304–310
146. Lingaiah N, Reddy KM, Nagaraju P, Sai Prasad PS, Wachs IE (2008) J Phy Chem C 112:8294–8300
147. Nagaraju P, Balaraju M, Reddy KM, Sai Prasad PS, Lingaiah N (2008) Catal Commun 9:1389–1393
148. Rao KTV, Rao PSN, Nagaraju P, Sai Prasad PS, Lingaiah N (2009) J Mol Catal A Chem 303:84–89
149. Nuemann R, Hagit M (1995) J Chem Soc Chem Commun 22:2277–2278
150. Mizuno N, Min JS, Taguchi A (2004) Chem Mater 16:2819–2825
151. Mazari T, Marchal CR, Hocine S, Salhi N, Rabia C (2009) J Nat Gas Chem 18:319–324
152. Hu J, Burns RC (2002) J Mol Catal A Chem 184:451–464
153. Palermo V, Romanelli GP, Vázquez PG (2013) J Mol Catal A Chem 373:142–150
154. Naumann R, Dahan M (1998) Polyhedron 17:3557–3564
155. Khenkin AM, Shimon LJW, Neumann R (2001) Eur J Inorg Chem 3:789–794
156. Knapp C, Ui T, Nagai K, Mizuno N (2001) Catal Today 71:111–119

157. Tundo P, Romanelli GP, Vázquezc PG, Aricò F (2010) Catal Commun 11:1181–1184
158. Karcz R, Pamin K, Poltowicz J, Haber J (2009) Catal Lett 132:159–167
159. Duarte TAG, Santos ICMS, Simões MMQ, Neves MGPMS, Cavaleiro AMV, Cavaleiro JAS
      (2013) Catal Lett 144:104–111

Chapter 2
Oxidation of Alcohols Catalyzed by Supported Undecamolybdophosphate

Abstract Synthesis and stabilization of undecamolybdophosphate (PMo_{11}) as well as its characterization was explained. Also, synthesis and characterization of series of heterogeneous catalysts comprising PMo_{11} and different supports (ZrO_2, Al_2O_3, MCM-41 and zeolite-Hβ) and their use as heterogeneous catalysts for solvent free oxidation of alcohols with molecular oxygen and TBHP as a radical initiator was described. The influence of different parameters on the conversion as well as the selectivity was investigated on oxidation of benzyl alcohol. Further, oxidation of various alcohols such as cyclopentanol, 1-hexanol and 1-octanol over supported PMo_{11} was presented under optimized conditions. All the catalysts are found to be efficient especially in achieving higher selective toward desire product and high turnover numbers. Regeneration study showed that the catalysts can be regenerated and reused without any significant loss in the catalytic activity.

Keywords Undecamolybdophosphate · Metal oxides · Porous silica · Oxidation · Alcohols

2.1 Undecamolybdophosphate Supported on to Metal Oxides (ZrO_2, Al_2O_3)

2.1.1 Experimental

All the chemicals used were of A. R. grade. Sodium molybdate, Disodium hydrogen phosphate, zirconium oxychloride, liquor ammonia, neutral active Al_2O_3 (Activity I–II, according to Brockmann), Nitric acid, acetone, benzyl alcohol, cyclopentanol, cyclohexanol, 1-hexanol, 1-octanol, tertiary butylhydroperoxide (70 % aq. TBHP) and dichloromethane were obtained from Merck and used as received.

© The Author(s) 2015
A. Patel and S. Pathan, *Polyoxomolybdates as Green Catalysts for Aerobic Oxidation*,
SpringerBriefs in Green Chemistry for Sustainability,
DOI 10.1007/978-3-319-12988-4_2

2.1.1.1 Synthesis of Na Salt of Undecamolybdophosphate (PMo$_{11}$) [1]

Sodium molybdate dihydrate (0.22 mol, 5.32 g) and anhydrous disodium hydrogen phosphate (0.02 mol, 0.28 g) were dissolved in 50–70 mL of conductivity water and heated to 80–90 °C followed by the addition of concentrated nitric acid in order to adjust the pH to 4.3. The volume was then reduced to half by evaporation and the heteropolyanion was separated by liquid–liquid extraction with 50–60 mL of acetone. The extraction was repeated until the acetone extract showed absence of NO$_3^-$ ions (ferrous sulfate test). The extracted sodium salt was dried in air. The obtained sodium salt of undecamolybdophosphate was designated as PMo$_{11}$.

2.1.1.2 Synthesis of Supported Catalysts

Synthesis of the support, ZrO$_2$
Hydrous zirconia was prepared by adding an aqueous ammonia solution to an aqueous solution of ZrOCl$_2$ · 8H$_2$O up to pH 8.5. The precipitates were aged at 100 °C over a water bath for 1 h, filtered, washed with conductivity water until chloride free water was obtained and dried at 100 °C for 10 h. The obtained material is designated as ZrO$_2$.

Supporting of PMo$_{11}$ onto ZrO$_2$ (PMo$_{11}$/ZrO$_2$) [1]
PMo$_{11}$ was supported on ZrO$_2$ by dry impregnating method. 1 g of ZrO$_2$ was impregnated with an aqueous solution of PMo$_{11}$ (0.1 g/10 mL of double distilled water). The water from the suspension was allowed to evaporate at 100 °C in an oven. Then the resulting mixture was dried at 100 °C with stirring for 10 h. The material thus obtained was designated as 10 % PMo$_{11}$/ZrO$_2$. Same procedure was followed for the synthesis of a series of supported PMo$_{11}$ containing 20–40 % PMo$_{11}$ (0.2–0.4 g/20–40 mL of conductivity water). The obtained materials were designated as 20 % PMo$_{11}$/ZrO$_2$, 30 % PMo$_{11}$/ZrO$_2$ and 40 % PMo$_{11}$/ZrO$_2$ respectively.

Supporting of PMo$_{11}$ onto Al$_2$O$_3$ (PMo$_{11}$/Al$_2$O$_3$) [2]
A series of catalysts containing 10–40 % PMo$_{11}$ were synthesized by impregnating 1 g of Al$_2$O$_3$ with an aqueous solution of PMo$_{11}$ (0.1–0.4 g in 10–40 mL of conductivity water) with stirring for 35 h and then dried at 100 °C for 10 h. The obtained materials were designated as 10 % PMo$_{11}$/Al$_2$O$_3$, 20 % PMo$_{11}$/Al$_2$O$_3$, 30 % PMo$_{11}$/Al$_2$O$_3$ and 40 % PMo$_{11}$/Al$_2$O$_3$.

2.1.1.3 Characterization

Characterization is a central aspect of catalyst development. The elucidation of the structures, compositions, and chemical properties of both the supports used in heterogeneous catalysis as well as the active species present on the surfaces of the supported catalysts is very important for a better understanding of the relationship between catalyst properties and catalytic performance [3]. Elemental analysis was carried out using JSM 5610 LV combined with INCA instrument for EDX-SEM. The total weight loss was calculated by the TGA method on the Mettler Toledo Star SW 7.01 upto 600 °C. FTIR spectra of the samples were recorded as the KBr pellet on the Perkin Elmer instrument. The spinning rate was 4–5 kHz. The BET specific surface area was calculated by using the standard Bruanuer, Emmett and Teller method on the basis of the adsorption data. Adsorption–Desorption isotherms of samples were recorded on a micromeritics ASAP 2010 surface area analyzer at −196 °C. The powder XRD pattern was obtained by using a Phillips Diffractometer (Model PW-1830). The conditions used were Cu Kα radiation (1.5417 Å).

2.1.1.4 Oxidation of Alcohols with Molecular Oxygen [4]

The catalytic activity was evaluated for oxidation of alcohols using molecular oxygen as an oxidant and *tert*-Butyl hydrogen peroxide as an initiator. Oxidation reaction was carried out in a batch type reactor operated under atmospheric pressure. In a typical reaction, measured amount of catalyst was added to a three necked flask containing alcohol at 90 °C. The reaction was started by bubbling O_2 into the liquid. The reactions were carried out by varying different parameters such as effect of % loading of PMo_{11}, reaction temperature, catalyst amount and reaction time.

After completion of the reaction, catalyst was removed and the product was extracted with dichloromethane. The product was dried with magnesium sulphate and analyzed on Gas Chromatograph (Nucon 5700 model) using BP-1 capillary column (30 m, 0.25 mm id). Product identification was done by comparison with authentic samples and finally by a combined Gas Chromatography Mass Spectrometer (Hewlett-Packard column) using HP-1 capillary column (30 m, 0.5 mm id) with EI and 70 eV ion source.

The conversion as well as selectivity was calculated on the basis of mole percent of substrates.

$$\text{Conversion}(\%) = \frac{(\text{initial mol}\,\%) - (\text{final mol}\,\%)}{(\text{initial mol}\,\%)}$$

$$\text{Selectivity}(\%) = \frac{\text{moles of product formed}}{\text{moles of substrate consumed}} \times 100$$

The turn over number (TON) was calculated using the following equation

$$\text{TON} = \frac{\text{moles of product}}{\text{moles of catalyst}}.$$

2.1.2 Result and Discussion

2.1.2.1 Characterization of Unsupported Undecamolybdophosphate

The PMo_{11} was isolated as the sodium salt after completion of the reaction and the remaining solution was filtered off. The filtrate was analyzed to estimate the amount of non reacted Mo [5]. The observed % of Mo in the filtrate was 0.5 %, which corresponds to loss of one equivalent of Mo from $H_3PMo_{12}O_{40}$. The observed values for the elemental analysis are in good agreement with the theoretical values indicating the formation of PMo_{11}. Anal. calc. (%): Na, 7.65; Mo, 50.12; P, 1.47; O, 39.52. Found (%): Na, 7.60; Mo, 49.99; P, 1.44; O, 39.92.

The TG-DTA curve of PMo_{11} is presented in Fig. 2.1. The TGA of PMo_{11} shows an initial weight loss of 16 % from 30 to 150 °C. This may be due to the removal of adsorbed water and water of crystallization. The final weight loss at around 235 °C indicates the decomposition of PMo_{11}.

DTA of PMo_{11} (Fig. 2.1) shows endothermic peaks at 80 and 140 °C, due to the loss of adsorbed water and water of crystallization respectively. In addition, DTA of PMo_{11} also shows a broad exothermic peak in the region of 270–305 °C. This may be due to the decomposition of PMo_{11}.

Number of water molecules was determined from the TGA curve using the following formula

$$18n = \frac{X(M + 18n)}{100}$$

where
n = number of water molecules
X = % loss from TGA
M = molecular weight of substance.

Fig. 2.1 TG-DTA of PMo_{11}

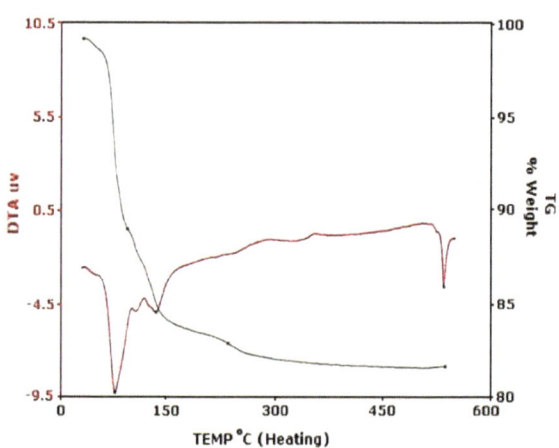

TEMP °C (Heating)

Fig. 2.2 FT-IR spectra for
a PMo$_{12}$ and **b** PMo$_{11}$

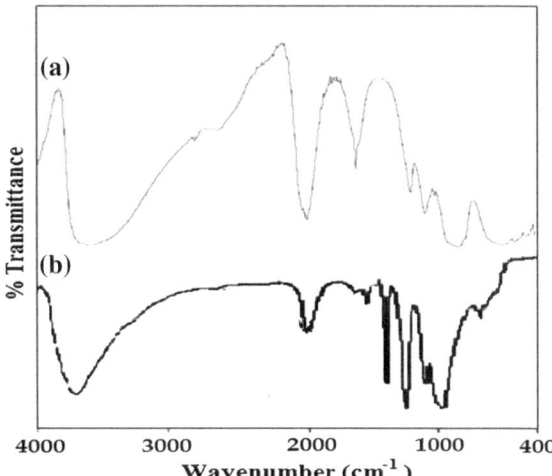

Based on studies the chemical formula of the isolated sodium salt was proposed as

$$Na_7[PMo_{11}O_{39}] \cdot 16H_2O$$

The FT-IR spectra of PMo$_{12}$ and PMo$_{11}$ are presented in Fig. 2.2. The bands at 1,070, 965, 870 and 790 cm^{-1} in parent Keggin PMo$_{12}$ attributed to asymmetric stretches of P–O$_a$, Mo–O$_t$, Mo–O–Mo.

The FT-IR spectra of PMo$_{11}$ shows P–O$_a$ band at 1,048 and 999 cm^{-1}. Observed splitting for P–O$_a$ band in PMo$_{11}$ as compare to that of PMo$_{12}$, indicates formation of lacunary structure with Keggin unit. The FT-IR spectra of PMo$_{11}$ also shows bands 935 and 906 cm^{-1} and 855 cm^{-1} attributed to asymmetric stretches of Mo–O$_t$ and Mo–O–Mo, respectively and are in good agreement with reported values [6]. The shifting in the band position may be due to formation of lacuna in synthesized material.

The ^{31}P chemical shift provides important information concerning the structure, composition and electronic states of these materials. ^{31}P MAS NMR of PMo$_{11}$ is presented in Fig. 2.3.

Van Veen [7] reported in acidic solution, phosphomolybdate was present in different type of species (e.g. $[P_2Mo_5O_{23}]^{6-}$, $[PMo_6O_{25}]^{9-}$, $[PMo_9O_{31} \cdot (OH)_3]^{6-}$, $[PMo_{10}O_{34}]^{3-}$, $[PMo_{11}O_{39}]^{7-}$ and $[PMo_{12}O_{40}]^{3-}$) in the acidified aqueous solution. The ^{31}P MAS NMR spectra shows single peak at 1.64 ppm corresponds to PMo$_{11}$ which also indicates absence of fragmentation of PMo$_{11}$.

The powder X-ray pattern for PMo$_{12}$ and PMo$_{11}$ is represented in Fig. 2.4. The powder XRD pattern of the isolated sodium salt indicates the semi-crystalline nature of PMo$_{11}$. PMo$_{11}$ shows the characteristic diffraction patterns with the typical 2θ value of 8–10° indicating the formation of lacunary molybdophosphate specie. For PMo$_{11}$, similar peaks with shifting as compare to that of PMo$_{12}$ were observed indicates presence of Keggin unit.

Fig. 2.3 ^{31}P MAS NMR of
PMo$_{11}$

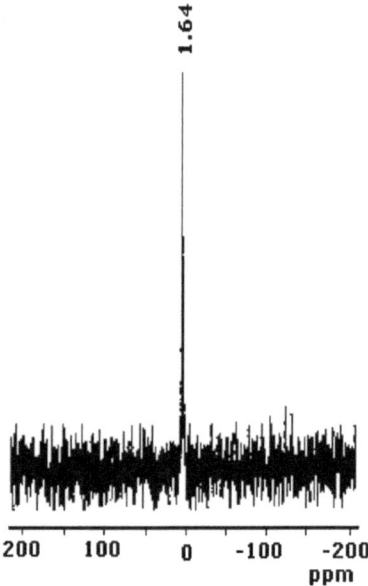

Fig. 2.4 Powder XRD
pattern of **a** PMo$_{12}$ and
b PMo$_{11}$

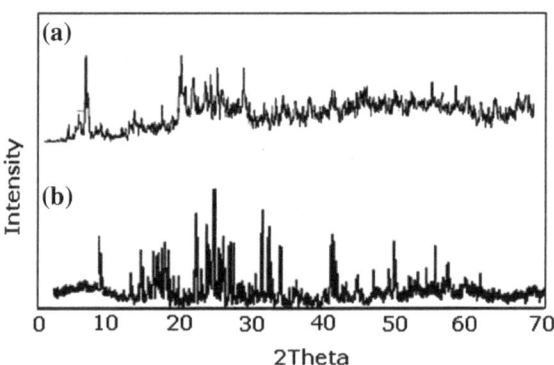

Thus the thermal, spectral as well as diffraction studies confirms the formation of PMo$_{11}$.

2.1.2.2 Characterization of Undecamolybdophosphate Supported on to Metal Oxides (ZrO$_2$, Al$_2$O$_3$)

Any leaching of the active species from the support makes the catalyst unattractive and hence it is necessary to study the stability as well as leaching of PMo$_{11}$ from the support. Heteropoly acids can be quantitatively characterized by the heteropoly blue colour, which is observed when it reacted with a mild reducing agent such as ascorbic acid. In the present study, this method was used for determining the leaching of PMo$_{11}$ from the support. Standard samples containing 1–5 % of PMo$_{11}$ in water were prepared. To 10 mL of the above samples, 1 mL of 10 % ascorbic acid was added. The mixture was diluted to 25 mL and the resultant solution was scanned at a λmax of 785 nm for its absorbance values. A standard calibration curve was obtained by plotting values of absorbance against % concentration. 1 g of supported catalyst with 10 mL conductivity water was refluxed for 18 h. Then 1 mL of the supernatant solution was treated with 10 % ascorbic acid. Development of blue colour was not observed indicating that there was no leaching. The same procedure was repeated with benzyl alcohol and the filtrate of the reaction mixture after completion of reaction in order to check the presence of any leached PMo$_{11}$. The absence of blue colour indicates no leaching of PMo$_{11}$.

The TG-DTA data of 20 % PMo$_{11}$/ZrO$_2$ and 20 % PMo$_{11}$/Al$_2$O$_3$ is presented in Table 2.1.

The TGA of 20 %-PMo$_{11}$/ZrO$_2$ and 20 %-PMo$_{11}$/Al$_2$O$_3$ (Table 2.1) shows 17 and 5 % weight loss up to 200 °C respectively, which is due to loss of adsorbed water. Catalysts do not show any weight loss up to 300 °C, indicating the synthesized catalysts is stable up to 300 °C. DTA of 20 % PMo$_{11}$/ZrO$_2$ and 20 % PMo$_{11}$/Al$_2$O$_3$ show an endothermic peak in range 80–140 °C which may be due to adsorbed water. Synthesized catalysts also show a broad exothermic peak in the region 285–325 and 255–425 °C, which may be due to decomposition of PMo$_{11}$ on the surface of the support. From the thermal study it is clearly seen that supported catalysts are stable upto 320 °C.

FT-IR was recorded to confirm the presence of reactive undegraded heteropolyanion species present on the surface of catalyst. The FT-IR data of 20 % PMo$_{11}$/ZrO$_2$ and 20 % PMo$_{11}$/Al$_2$O$_3$ is presented in Table 2.2. As described earlier (Fig. 2.2), FT-IR of PMo$_{11}$ shows a band at 1,048 and 999 cm^{-1}, 935, 906 and 855 cm^{-1} attributed to asymmetric stretches of P–O, Mo–O$_t$ and Mo–O–Mo, respectively. The synthesized catalysts retained similar bands of Keggin unit i.e. bands at 1,039, 990 and 910 cm^{-1} for 20 % PMo$_{11}$/ZrO$_2$ as well as bands at 1,052, 1,010 and 914 cm^{-1} for 20 % PMo$_{11}$/Al$_2$O$_3$ correspond to asymmetric stretching of P–O and Mo–O–Mo, respectively. The shifting of bands as well as the disappearance of the Mo–Ot band (935 cm^{-1}) for the supported catalysts may be due to interaction of terminal oxygen of PMo$_{11}$ with surface hydrogen of support.

As described earlier, ^{31}P MAS NMR spectra of PMo$_{11}$ (Fig. 2.3) show a single peak at 1.64 ppm. While, ^{31}P MAS NMR for 20 % PMo$_{11}$/ZrO$_2$ and 20 % PMo$_{11}$/Al$_2$O$_3$ (Fig. 2.5) reveals resonance at −2.42 and −3.47 ppm respectively.

Table 2.1 TG-DTA data PMo$_{11}$, 20 % PMo$_{11}$/ZrO$_2$ and 20 % PMo$_{11}$/Al$_2$O$_3$

Catalyst	TGA (% Weight loss up to 200 °C)	DTA (°C)	
		Endothermic	Exothermic
20 % PMo$_{11}$/ZrO$_2$	17 (continues)	80	285–325
20 % PMo$_{11}$/Al$_2$O$_3$	9 (continues)	80, 140	255–425

Table 2.2 FT-IR data PMo$_{11}$, 20 % PMo$_{11}$/ZrO$_2$ and 20 % PMo$_{11}$/Al$_2$O$_3$

Catalyst	FT-IR frequency (cm^{-1})		
	P–O	Mo–O$_t$	Mo–O$_{b,c}$–Mo
20 % PMo$_{11}$/ZrO$_2$	1,039, 990	–	910
20 % PMo$_{11}$/Al$_2$O$_3$	1,052, 1,010	–	914

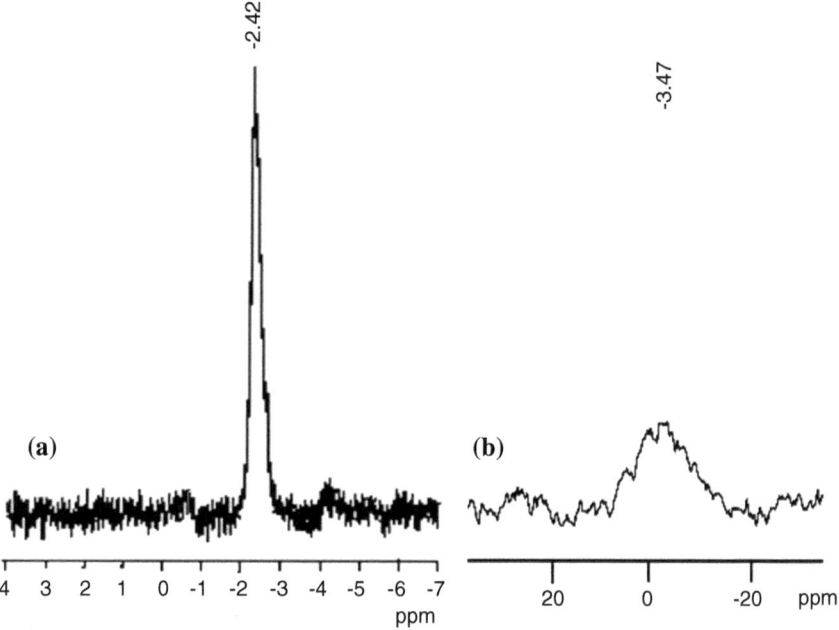

Fig. 2.5 ^{31}P MAS NMR of **a** 20 % PMo$_{11}$/ZrO$_2$ and **b** 20 % PMo$_{11}$/Al$_2$O$_3$

It is reported that the lower shift, i.e. deshielding, increases as the degree of adsorption and degree of fragmentation increases [8]. In the present cases, the observed chemical shift, shielding as compare to that of PMo$_{11}$, indicates presence of chemical interaction between PMo$_{11}$ and supports, rather than simple adsorption. Also, a single peak for both the catalysts indicated that no fragmentation of PMo$_{11}$ species takes place after addition to the support. Thus it can be concluded that PMo$_{11}$ remains intact on surface of supports, after supporting on to supports.

The difference in nature of the signal as well as values of the NMR shift for 20 % PMo$_{11}$/ZrO$_2$ and 20 % PMo$_{11}$/Al$_2$O$_3$ may be due to the different nature of the supports. It is known that ZrO$_2$ is an acidic support and it has more surface hydroxyl groups as compare to that of Al$_2$O$_3$ for strong interaction through hydrogen bonding with the terminal oxygens of PMo$_{11}$. Thus, more shielding is observed for Al$_2$O$_3$. Thus the obtained results are in good agreement with the proposed explanation.

The values of surface area for both the series of catalysts are shown in Table 2.3. It is seen from Table 2.3 that, initially the value for surface area increases with increase in loading from 10 to 20 % while it decreases from 20 to 40 %. This may be due to the formation of multilayers of active species, PMo$_{11}$, onto support surface, which may cause blocking/stabilization of active sites on the monolayer.

The XRD pattern of 20 % PMo$_{11}$/ZrO$_2$ (Fig. 2.6a) and 20 % PMo$_{11}$/Al$_2$O$_3$ (Fig. 2.6b) shows the amorphous nature of the materials indicating that the crystallinity of the PMo$_{11}$ is lost after supporting.

Further, it does not show any diffraction lines of lacunary PMo$_{11}$ indicating a very high dispersion of solute as a non-crystalline form on the surface of the supports.

In order to compare the surface morphology of both the supported catalysts, the SEM pictures were taken at same magnification. The scanning electron microscopy (SEM) images of PMo$_{11}$, 20 % PMo$_{11}$/ZrO$_2$ and 20 % PMo$_{11}$/Al$_2$O$_3$ at a magnification of 100× are reported in Fig. 2.7.

Figure 2.7a shows the semi-crystalline nature of PMo$_{11}$. SEM images of 20 % PMo$_{11}$/ZrO$_2$ and 20 % PMo$_{11}$/Al$_2$O$_3$ (Fig. 2.7b, c) showed a uniform dispersion of PMo$_{11}$ in a non-crystalline form on the surface of the support.

Table 2.3 Surface area

Catalysts	Surface area (m^2 g^{-1})
ZrO$_2$	170
10 % PMo$_{11}$/ZrO$_2$	191
20 % PMo$_{11}$/ZrO$_2$	197
30 % PMo$_{11}$/ZrO$_2$	188
40 % PMo$_{11}$/ZrO$_2$	187
Al$_2$O$_3$	81.0
10 % PMo$_{11}$/Al$_2$O$_3$	99.7
20 % PMo$_{11}$/Al$_2$O$_3$	101.8
30 % PMo$_{11}$/Al$_2$O$_3$	72.9
40 % PMo$_{11}$/Al$_2$O$_3$	56.2

Fig. 2.6 Powder XRD pattern of **a** 20 % PMo$_{11}$/ZrO$_2$ and **b** 20 % PMo$_{11}$/Al$_2$O$_3$

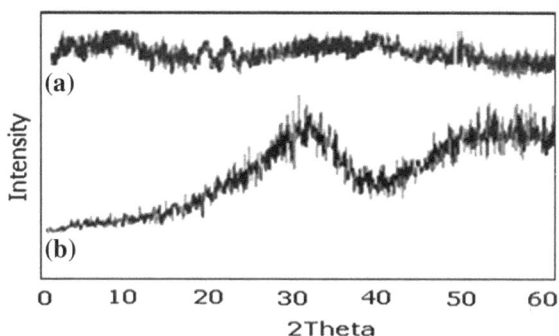

Fig. 2.7 SEM images of **a** PMo_{11} and **b** 20 % PMo_{11}/ZrO_2, **c** 20 % PMo_{11}/Al_2O_3

2.1.2.3 Catalytic Activity

A detail study was carried out on oxidation of benzyl alcohol to optimize the conditions. To ensure the catalytic activity, all reactions were carried out without catalyst and no oxidation takes place. The reaction was carried out with 25 mg of PMo_{11}/ZrO_2 and PMo_{11}/Al_2O_3 for 24 h at 90 °C. Generally, benzyl alcohol on oxidation gives benzaldehyde and benzoic acid. However, benzaldehyde was found as the major oxidation product in the present case (Scheme 2.1).

Scheme 2.1 Oxidation of benzyl alcohol

Benzyl alcohol — Oxidant / Catalyst → Benzaldehyde (Major) → Benzoic acid

The oxidation of benzyl alcohol was carried out using 25 mg of catalysts for 24 h at 90 °C are presented in Fig. 2.8.

From Fig. 2.8 it is clear that, increase in the conversion was observed with increase in the % loading of PMo$_{11}$ from 10 to 20 %. Further, increase in % loading from 20 to 40 %, no significant increase in conversion was observed. This may be due to at higher % loading, the particles may agglomerate on the surface, results in reduce accessibility to the active sites. Thus, loading of PMo$_{11}$ on the supports was fixed at 20 % and detail studies were carried out over 20 % PMo$_{11}$/ZrO$_2$ and 20 % PMo$_{11}$/Al$_2$O$_3$.

In order to determine the optimum temperature the reaction was investigated at four different temperatures 80, 90 and 100 °C, using both the catalysts keeping other parameters fixed (amount 25 mg, time 24 h). The results for the same are presented in Fig. 2.9.

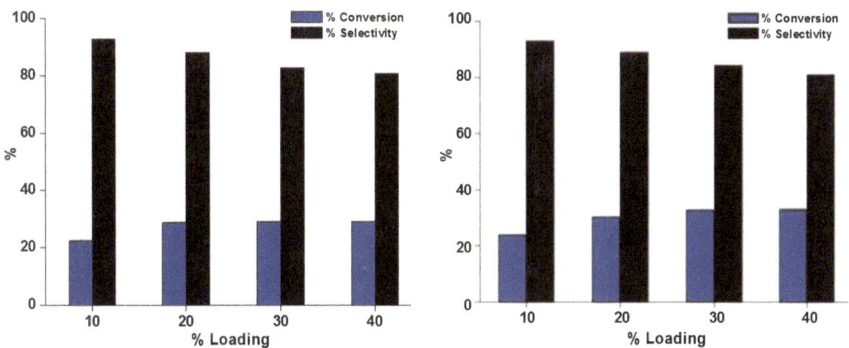

Fig. 2.8 % Conversion is based on benzyl alcohol; benzyl alcohol = 100 mmol, TBHP = 0.2 %, temp = 90 °C, time = 24 h, amount of catalyst = 25 mg, PMo$_{11}$/ZrO$_2$ (*left*) and PMo$_{11}$/Al$_2$O$_3$ (*right*)

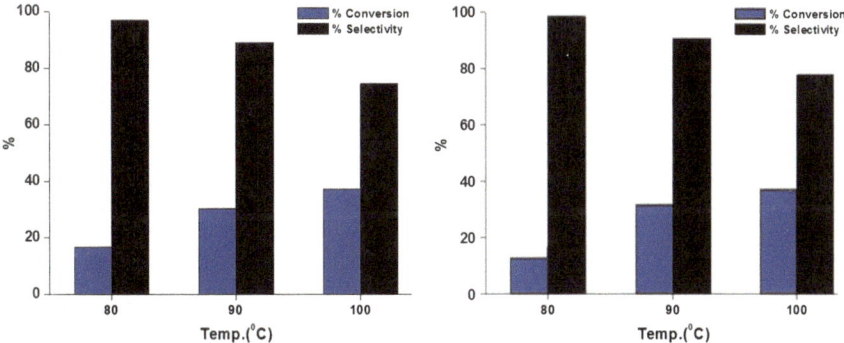

Fig. 2.9 % Conversion is based on benzyl alcohol; benzyl alcohol = 100 mmol, TBHP = 0.2 %, time = 24 h, amount of catalyst = 25 mg, PMo$_{11}$/ZrO$_2$ (*left*) and PMo$_{11}$/Al$_2$O$_3$ (*right*)

The results show that conversion increased with increasing temperature. At the same time, on increasing temperature from 90 to 100 °C, selectivity of benzaldehyde was decrease for both catalysts, this is due to over oxidation of benzaldehyde to benzoic acid at elevated temperature. So temperature of 90 °C was fixed for the optimum conversion of benzyl alcohol as well selectivity of desire product.

The effect of the catalyst amount on the conversion of benzyl alcohol over both the catalysts is represented in Fig. 2.10.

The catalyst amount was varied from 15 to 50 mg. It is seen from this figure that the activity increases initially up to 25 mg and then becomes constant with further increase in the amount of catalyst. However, decrease in selectivity was observed. In the present case, water is formed as a by-product which is polar in nature. In presence of polar molecules, heteropolyacids exhibit pseudoliquid behaviour [9] in which catalytic activity is directly proportional to the active amount of the catalyst. So the same observation is expected in case of LPOMs. Further, increase in the amount blocks/stabilizes the active sites. Hence, no change in conversion is expected. The obtained results are in good agreement with the proposed explanation.

The effect of reaction time is shown in Fig. 2.11. It is seen from figure that initially with increase in reaction time, the conversion also increases. After some time, the conversion attains stability. This may be due to the fact that the activation of the catalyst as well as the attainment of equilibrium requires time.

Once the equilibrium is attained, the conversion becomes almost constant. But at the same time, selectivity of benzaldehyde decreases. This may be due to over oxidation of benzaldehyde to benzoic acid. Thus, reaction time at 24 h was optimized for optimum conversion of benzyl alcohol as well selectivity of benzaldehyde.

The optimum conditions for 23.7 % conversion with 92.3 % selectivity for benzaldehyde with 20 %-PMo_{11}/ZrO_2 and 22.5 % conversion with 94.8 % selectivity for benzaldehyde with 20 %-PMo_{11}/Al_2O_3 are, catalyst amount = 25 mg, temperature = 90 °C, time = 24 h.

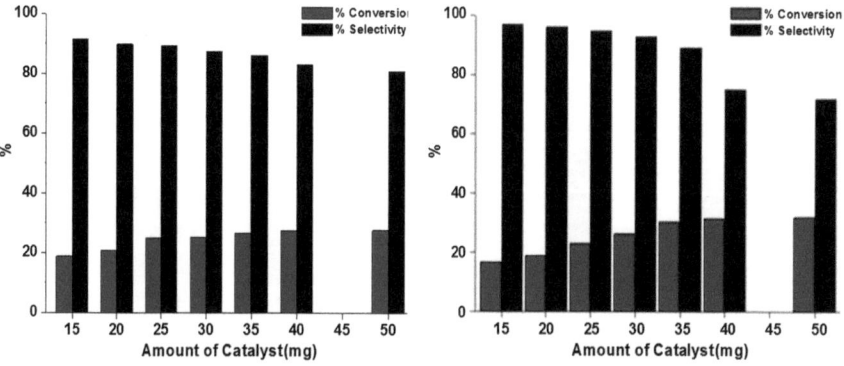

Fig. 2.10 % Conversion is based on benzyl alcohol; benzyl alcohol = 100 mmol, TBHP = 0.2 %, temp = 90 °C, time = 24 h, PMo_{11}/ZrO_2 (*left*) and PMo_{11}/Al_2O_3 (*right*)

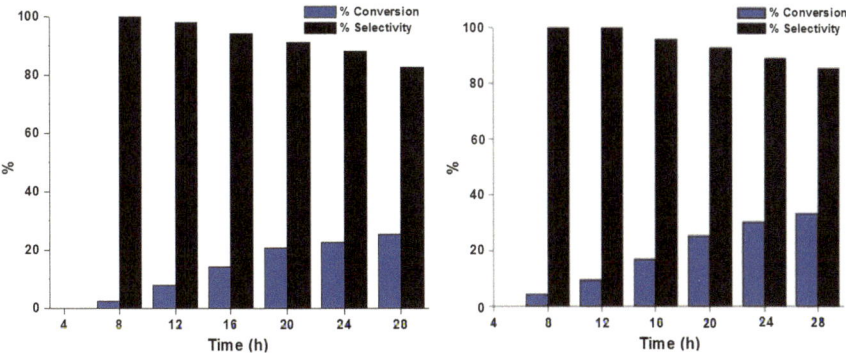

Fig. 2.11 % Conversion is based on benzyl alcohol; benzyl alcohol = 100 mmol, TBHP = 0.2 %, temp = 90 °C, amount of catalyst = 25 mg, PMo$_{11}$/ZrO$_2$ (*left*) and PMo$_{11}$/Al$_2$O$_3$ (*right*)

In order to see scope and limitations of present catalytic systems, oxidation of various alcohols was carried out with supported catalysts under optimized conditions and results are shown in Table 2.4. It was observed from Table 2.4 that, oxidation of secondary alcohol is easier as compare to primary alcohols.

The observed trend is in good agreement with reported in art [10]. In all cases, very high TON is observed. It is known that, oxidation of long chain alcohols (C$_8$ and onwards) is still challenging task because of lower reactivity [11] and thus present catalytic system is also not applicable to less reactive long chain primary alcohol such as 1-octanol. For the present catalytic system, the reactivity of the alcohols was found in the order primary < cyclic secondary < aromatic.

Table 2.4 Oxidation of various alcohols with O$_2$, under optimized conditions

Alcohols	Conversion (%)	Products	Selectivity (%)	TON
[a]Benzyl alcohol	23.7	Benzaldehyde	92.3	11,814
		Benzoic acid	7.7	
[a]Cyclopentanol	22.7	Cyclopentanone	>99	11,609
[a]Cyclohexanol	22.4	Cyclohexanone	>99	11,456
[a]1-Hexanol	9.1	1-Hexanal	>99	4,654
[a]1-Octanol	–	–	–	–
[b]Benzyl alcohol	22.5	Benzaldehyde	94.8	11,216
		Benzoic acid	5.2	
[b]Cyclopentanol	22.2	Cyclopentanone	>99	11,353
[b]Cyclohexanol	21.6	Cyclohexanone	>99	11,046
[b]1-Hexanol	8.2	1-Hexanal	>99	4,194
[b]1-Octanol	–	–	–	–

% Conversion is based on alcohol; alcohol = 100 mmol, TBHP = 0.2 %, temp = 90 °C, time = 24 h, amount of catalyst = 25 mg
[a] (PMo$_{11}$)$_2$/ZrO$_2$
[b] (PMo$_{11}$)$_2$/Al$_2$O$_3$

It is well know that support does not play always merely a mechanical role but it can also modify the catalytic properties of the catalysts. So, in order to see the effect of support, comparison of activity of 20 % PMo_{11}/ZrO_2 and 20 % PMo_{11}/Al_2O_3 was done for oxidation of benzyl alcohol under optimized conditions and results are presented in Table 2.4. Generally, oxidation of benzyl alcohol gives benzaldehyde (major), with over oxidation product benzoic acid (minor). These types of over oxidation reactions are directly promoted by acidity of the catalyst. So, observed result i.e. lower selectivity of benzaldehyde in case of PMo_{11}/ZrO_2 is attributed to acidity of ZrO_2.

2.1.2.4 Heterogeneity Test

Heterogeneity test was carried out for the oxidation of benzyl alcohol (Fig. 2.12) over PMo_{11}/ZrO_2 as examples.

For the rigorous proof of heterogeneity, a test [12] was carried out by filtering catalyst from the reaction mixture at 90 °C after 16 h and the filtrate was allowed to react up to 24 h. Reaction mixture of 16 h and the filtrate were analyzed by gas chromatography. Similar test was carried for PMo_{11}/Al_2O_3. No change in the % conversion as well as % selectivity was found indicating the present catalysts fall into category C [12] i.e., active species does not leach and the observed catalysis is truly heterogeneous in nature.

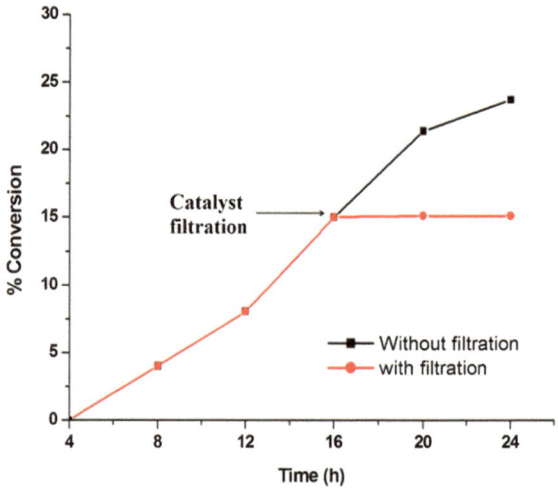

Fig. 2.12 % Conversion is based on benzyl alcohol; benzyl alcohol = 100 mmol, O_2, TBHP = 0.2 %, amount of 20 % PMo_{11}/ZrO_2 = 25 mg; temperature 90 °C

Table 2.5 Oxidation of benzyl alcohol with fresh and regenerated catalysts

Cycle	Conversion (%)	Selectivity (%)
		Benzaldehyde
Fresh	23.1[a]/23.0[b]	92.3[a]/94.6[b]
1	22.9[a]/22.9[b]	92.2[a]/95.0[b]
2	22.9[a]/22.9[b]	92.2[a]/94.9[b]
3	22.9[a]/22.5[b]	92.0[a]/94.2[b]
4	22.7[a]/22.6[b]	92.0[a]/94.2[b]

% Conversion is based on benzyl alcohol; benzyl alcohol = 100 mmol, TBHP = 0.2 %, temp = 90 °C, time = 24 h, amount of catalyst = 25 mg
[a] 20 % PMo$_{11}$/ZrO$_2$
[b] 20 % PMo$_{11}$/Al$_2$O$_3$

2.1.2.5 Catalytic Activity of Regenerated Catalysts

The catalyst was recycled in order to test for its activity as well as its stability. The catalysts remain insoluble under the present reaction conditions. The leaching of Mo from catalyst support was confirmed by carrying out an analysis of the used catalyst (EDS) as well as the product mixtures (AAS). The analysis of the used catalyst did not show an appreciable loss in the Mo content as compared to the fresh catalyst. The analysis of the product mixtures showed that if any Mo was present it was there in an amount below the detection limit, which corresponded to less than 1 ppm. These observations strongly suggest that the present catalyst is truly heterogeneous in nature.

Catalysts were separated easily by simple filtration followed by washing with dichloromethane and dried at 100 °C. Oxidation reaction was carried out with the regenerated catalysts, under the optimized conditions. The data for the catalytic activity is represented in Table 2.5. It is seen from the table that there was no change in selectivity, however, a little decrease in conversion was observed. This shows that the catalysts are stable, regenerated and reused successfully up to 4 cycles.

2.1.2.6 Characterization of Regenerated Catalysts

20 % PMo$_{11}$/ZrO$_2$ and 20 % PMo$_{11}$/Al$_2$O$_3$ were regenerated and characterized for FT-IR in order to confirm the retention of the catalyst structure, after the completion of the reaction. The FT-IR data for the fresh as well as the regenerated catalysts are represented in Fig. 2.13. No appreciable shift in the FT-IR band position of the regenerated catalyst indicates the retention of Keggin type PMo$_{11}$ onto supports.

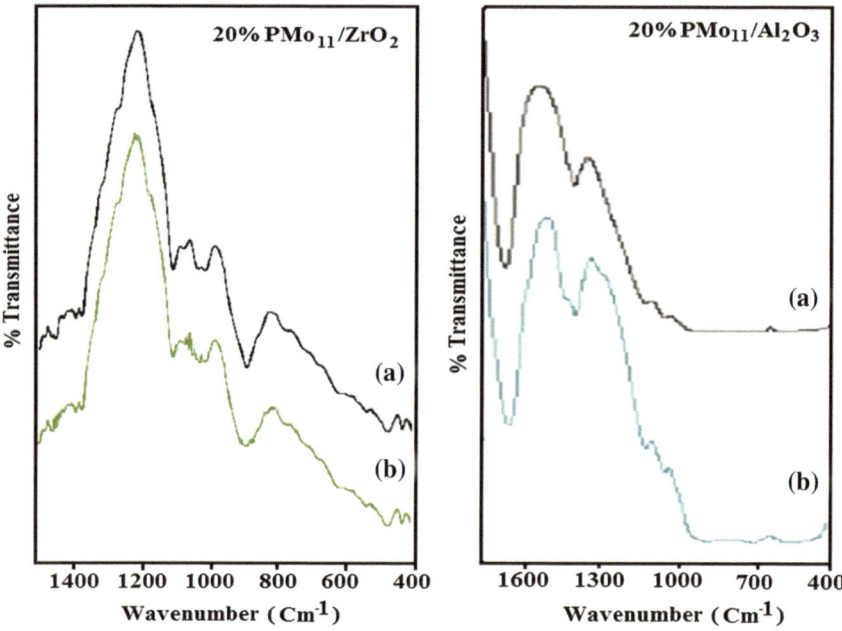

Fig. 2.13 FT-IR spectra of **a** fresh and **b** recycled (fourth cycle) catalysts

2.2 Undecamolybdophosphate Anchored to Porous Supports (Zeolite Hβ, MCM-41)

2.2.1 Experimental

2.2.1.1 Materials

All the chemicals used were of A. R. grade. CTAB (cetyltrimethyl ammonium bromide), TEOS (tetraethyl orthosilicate), benzyl alcohol, cyclopentanol, cyclo-hexanol, 1-hexanol, 1-octanol, 70 % aq. TBHP, NaOH and dichloromethane were obtained from Merck and used as received. The sodium form of zeolite β (Naβ) with Si/Al ratio 10 was purchased from Zeolites and Allied Products, Bombay, India, and used as received.

2.2.1.2 Synthesis of the Supports (MCM-41, Hβ) [13]

CTAB was added to a very dilute solution of NaOH with stirring at 60 °C. When the solution became homogeneous, TEOS was added dropwise, and the obtained gel was aged for 2 h. The resulting product was filtered, washed with distilled water, and then dried at room temperature. The obtained material was calcined at 555 °C in air for 5 h and designated as MCM-41.

Treatment of the Support (zeolite-Hβ)

Naβ was converted in to the NH_4^+ form by conventional ion exchange method [14] using a 10 wt%, 1 M NH_4Cl aqueous solution. The resulting NH_4^+ type zeolite was further converted to H^+ type by calcination in air at 550 °C for 6 h.

2.2.1.3 Synthesis of Supported Catalysts [15]

Supporting of PMo₁₁ to MCM-41 and Hβ

A series of catalysts containing 10–40 % of PMo_{11} supported to support (MCM-41 and Hβ) were synthesized by impregnation method. One gram of support was impregnated with an aqueous solution of PMo_{11} (0.1/10–0.4/40 g/mL of double distilled water) and dried at 100 °C for 10 h. The obtained materials were designated as 10 % PMo_{11}/MCM-41, 20 % PMo_{11}/MCM-41, 30 % PMo_{11}/MCM-41, and 40 % PMo_{11}/MCM-41, respectively, as well as 10 % PMo_{11}/Hβ, 20 % PMo_{11}/Hβ, 30 % PMo_{11}/Hβ, and 40 % PMo_{11}/Hβ, respectively.

Characterization specifications and catalytic reaction is similar as mentioned in Sects. 2.1.1.3 and 2.1.1.4.

2.2.2 Result and Discussion

2.2.2.1 Characterization of Undecamolybdophosphate Anchored to Porous Supports (Zeolite-Hβ, MCM-41)

TGA curves of zeolite support and catalysts are shown in Fig. 2.14. The TGA of PMo_{11} (Fig. 2.14a) shows the initial weight loss of 16 % from 30 to 200 °C. This may be due to the removal of adsorbed water molecules. The final weight loss of 1.8 % at around 275 °C indicates the loss of water of crystallisation. A unique weight loss of 13–15 % was observed up to 250 °C for zeolite support (Fig. 2.14b), which is attributed to desorption of physically adsorbed water. No further weight loss was observed beyond 250 °C which indicates zeolite Hβ retains its framework structure up to 600 °C. TGA of MCM-41 (Fig. 2.14c) shows initial weight loss of 6.14 % at 100 °C. This may be due to the loss of adsorbed water molecules. The final 7.92 % weight loss above 450 °C may be due to the condensation of silanol groups to form siloxane bonds. After that, the absence of any weight loss indicates that support is stable up to 550 °C. The TGA of 30 % PMo_{11}/Hβ (Fig. 2.14d) shows initial weight loss of 11.2 % up to 150 °C which may be due to the removal of adsorbed water molecules. No significant weight loss occurs up to 350 °C (2.6 %), which indicates the stability of the catalyst up to 350 °C. The removal of the Mo–O moiety from the parent PMo_{12} leads to the thermally less stable material PMo_{11}.

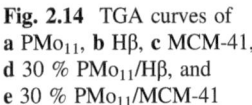

Fig. 2.14 TGA curves of
a PMo$_{11}$, **b** Hβ, **c** MCM-41,
d 30 % PMo$_{11}$/Hβ, and
e 30 % PMo$_{11}$/MCM-41

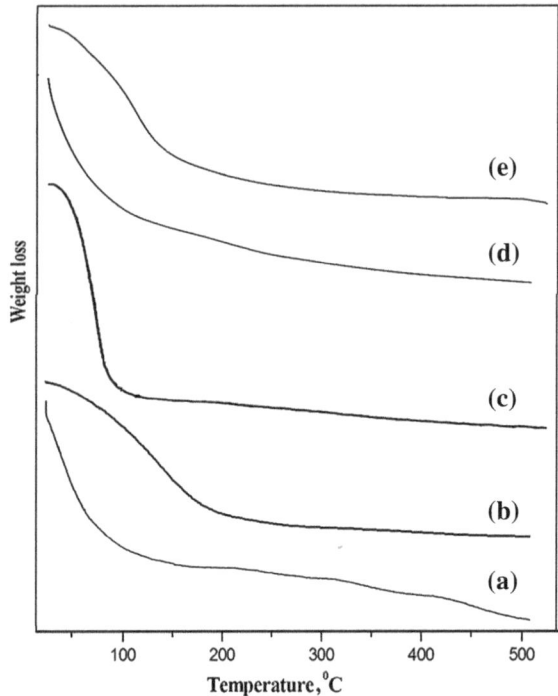

The thermal stability values of PMo$_{11}$ and supported PMo$_{11}$ are 305 and 350 °C respectively. The TGA of 30 % PMo$_{11}$/MCM-41 (Fig. 2.14e) shows initial weight loss up to 150 °C may be due to the removal of adsorbed water molecules. No significant loss occurs up to 350 °C, which indicates the stability of the catalyst up to 350 °C.

The FT-IR spectra of MCM-41, Hβ, PMo$_{11}$ and 30 % PMo$_{11}$/Hβ and 30 % PMo$_{11}$/MCM-41 are presented in Fig. 2.15. The FT-IR spectrum of MCM-41 (Fig. 2.15b) shows a broad band around 1,300–1,000 cm^{-1} corresponding to v_{as} (Si–O–Si). The bands at 801 and 458 cm^{-1} are due to symmetric stretching and bending vibration of Si–O–Si, respectively. The band at 966 cm^{-1} corresponds to v_s (Si–OH). The bands at 1048 and 999 cm^{-1}, 935 and 906 cm^{-1} and 855 cm^{-1} in parent PMo$_{11}$ (Fig. 2.15a) are attributed to asymmetric stretching of P–O, Mo–Ot and Mo–O–Mo respectively. The spectrum of 30 % PMo$_{11}$/MCM-41 (Fig. 2.15c) shows bands at 965 and 918 cm^{-1} assigned to P–O and Mo = O stretches, respectively. The shift in the bands is an indication of strong chemical interaction between PMo$_{11}$ and the silanol groups of MCM-41.

The FT-IR spectra for Hβ and 30 % PMo$_{11}$/Hβ (Fig. 2.15) shows large and broad peak appearing in range 1,060–1,090 cm^{-1} due to asymmetric stretching vibration O–T–O (v_{asym}), which is sensitive to the silicon and aluminum contents in the zeolite framework. A band at 455 cm^{-1} is characteristic of the pore opening. The shifting and broadening of bands as well as disappearance of Mo–Ot band

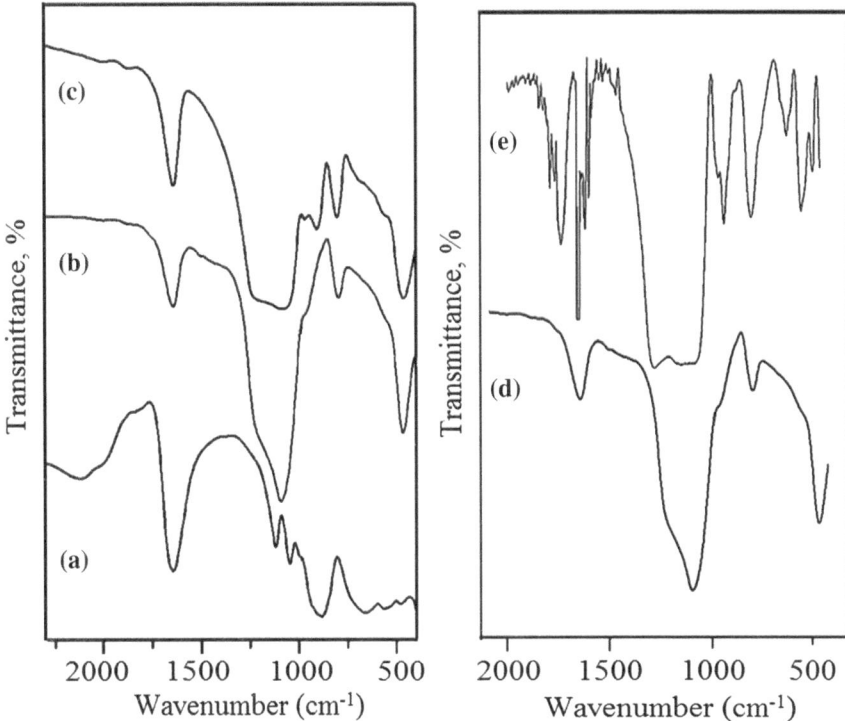

Fig. 2.15 FT-IR of **a** PMo$_{11}$, **b** MCM-41, **c** 30 % PMo$_{11}$/MCM-41, **d** Hβ and **e** 30 % PMo$_{11}$/Hβ

(935 cm^{-1}) may be due to strong chemical interaction of terminal oxygen of PMo$_{11}$ with Hβ, which was further confirmed by FT-Raman analysis.

FT-Raman spectra of PMo$_{11}$ and both the catalysts are displayed in Fig. 2.16. The FT-Raman spectrum of PMo$_{11}$ shows typical bands at 947 (v_s(Mo–O$_d$)), 932 (v_{as}(Mo–O$_d$)), 891 and 550 (v_{as}(Mo–Ob–Mo)), 355 (v_{as}(O$_a$–P–O$_a$)) and 217 (v_s(Mo–O$_a$)), where O$_a$, O$_b$, O$_c$, and O$_d$ are attributed to the oxygen atoms connected to phosphorus, to oxygen atoms bridging two molybdenums (from two different triads for O$_b$ and from the same triad for O$_c$), and to the terminal oxygen Mo = O, respectively. The FT-Raman spectrum of 30 % PMo$_{11}$/MCM-41 shows the retention of all the characteristic bands at 916 (v_s(Mo–O$_d$)), 876 (v_{as}(Mo–O$_d$)), 793 and 548 (v_{as}(Mo–O$_b$–Mo)), 323 (v_{as}(O$_a$–P–O$_a$)) and 140 (vs(Mo–O$_a$)), which indicates that the structure of PMo$_{11}$ has been retained after anchoring to the support. The FT-Raman spectrum of 30 % PMo$_{11}$/Hβ (Fig. 2.16) shows the retention of all the characteristic bands of PMo$_{11}$ which indicates that the structure of PMo$_{11}$ has been retained after anchoring to the support. However, a large shift was observed for all characteristic bands indicating the presence of very strong interaction between the oxygen of PMo$_{11}$ and Hβ.

Fig. 2.16 FT-Raman spectra
of **a** PMo$_{11}$, **b** 30 %
PMo$_{11}$/MCM-41, and
c 30 % PMo$_{11}$/Hβ

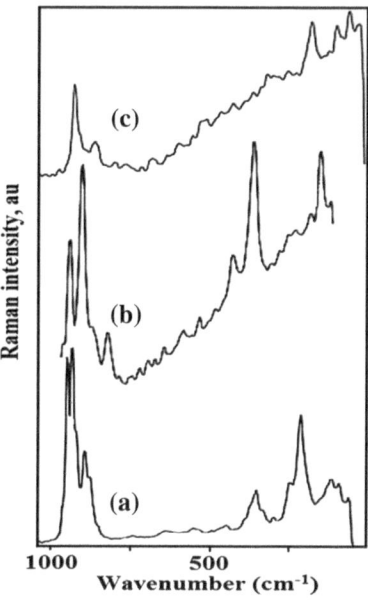

The decrease in the specific surface area for 30 % PMo$_{11}$/MCM-41 (485 m^2 g^{-1}) as compared to that of MCM-41 (659 m^2 g^{-1}) is as expected and is the first indication of chemical interaction between available surface oxygen of PMo$_{11}$ and the proton of the silanol group of MCM-41. Figure 2.17 shows the nitrogen adsorption–desorption isotherms and BET pore size distribution curves for Hβ and 30 % PMo$_{11}$/Hβ. Both Specific surface area and pore diameter strongly decreased for PMo$_{11}$-containing Hβ relative to the starting support. The decrease in the specific surface area for 30 % PMo$_{11}$/Hβ (362 m^2 g^{-1}) as compared to that of Hβ (587.2 m^2 g^{-1}) is as expected and is the first indication of chemical interaction between available surface oxygen of PMo$_{11}$ and Hβ. The nitrogen adsorption

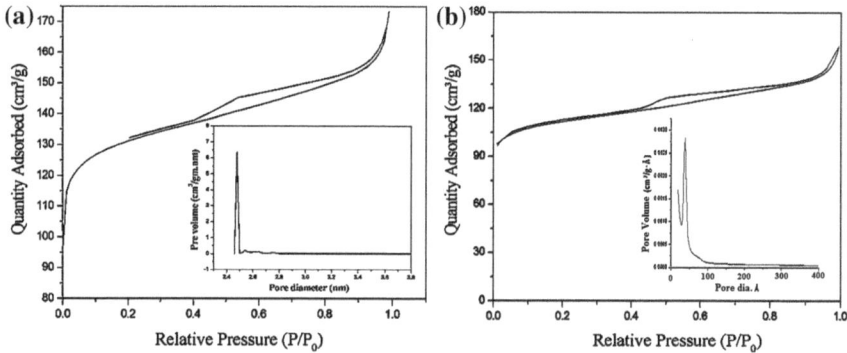

Fig. 2.17 Nitrogen sorption and pore size distribution of **a** Hβ and **b** 30 % PMo$_{11}$/Hβ

isotherms of support and catalyst are displayed in Fig. 2.17. Both samples showed Type (IV) pattern with three stages: monolayer adsorption of nitrogen on the walls of mesopores (P/Po < 0.4), the part characterized by a steep increase in adsorption due to capillary condensation in mesopores with hysteresis (P/Po = 0.4–0.8), and multilayer adsorption on the outer surface of the particles. It was observed that pore diameter was decreased after anchoring PMo_{11} on to the support.

The FT-IR, FT-Raman, BET shows that PMo_{11} remains intact even after anchoring and there exists a strong chemical interaction of PMo_{11} with the support.

2.2.2.2 Catalytic Activity

In the present work, a detailed catalytic study was carried out over 30 % PMo_{11}/Hβ for the oxidation of benzyl alcohol by varying different parameters like % loading, catalyst amount, reaction time and temperature to optimize the conditions for the maximum conversion.

Figure 2.18 shows the effect of % loading of PMo_{11} on the oxidation of benzyl alcohol using molecular oxygen. The conversion increases with increase in the % loading of PMo_{11} from 10 to 20 %. There is a drastic increase in the conversion on increasing loading from 20 to 30 %. The conversion reaches 25.5 % for 30 % loaded catalyst and further increase in the loading to 40 % does not affect the conversion significantly. This is because, at higher % loading, the active species agglomerate on the surface of the support resulting in low accessibility to the active sites. It is also seen that at higher loading the selectivity for benzaldehyde decreases. Hence, 30 % loading was selected for further study.

The amount of catalyst has a significant effect on the oxidation of benzyl alcohol. Reaction was carried out by taking catalyst amount in the range 15–30 mg (Fig. 2.19). Initially, on increasing catalyst amount from 10 to 25 mg, the conversion increases sharply. Further increase in the catalyst amount does not increase

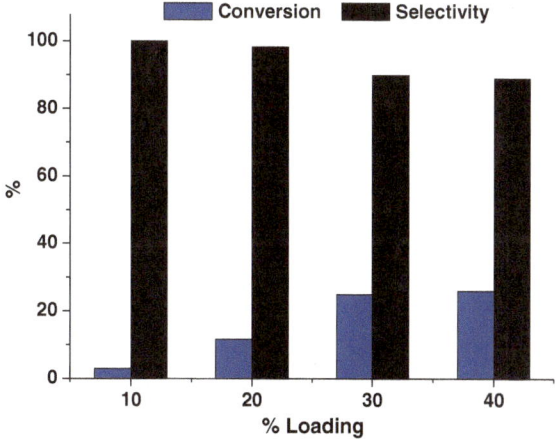

Fig. 2.18 % Conversion is based on benzyl alcohol; benzyl alcohol = 100 mmol, TBHP = 0.2 mmol, amount of catalyst = 25 mg, time = 24 h, temp = 90 °C

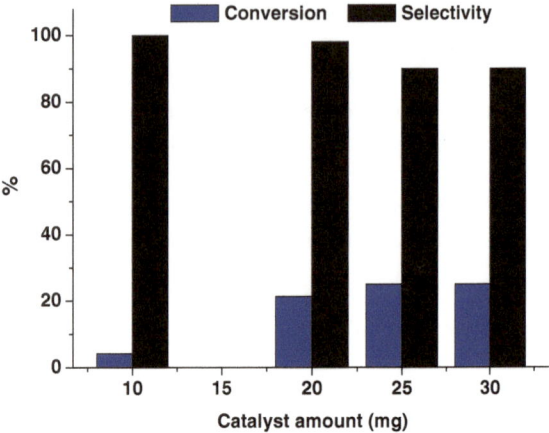

Fig. 2.19 % Conversion is based on benzyl alcohol; benzyl alcohol = 100 mmol, TBHP = 0.2 mmol, time = 24 h, temp = 90 °C

the conversion very significantly. Therefore, 25 mg amount of catalyst has been considered as optimum for the maximum conversion.

The effect of reaction time on the selective oxidation was carried out by monitoring % conversion at different time intervals. It is seen from Fig. 2.20 that with increase in reaction time, the % conversion also increases. Initial conversion of benzyl alcohol increased with the reaction time. This is due to the reason that more time is required for the formation of reactive intermediate (substrate + catalyst) which finally converts into the products. It was seen that 25.5 % conversion was observed at 24 h; when the reaction was allowed to continue for 25 h, no significant conversion was observed, but selectivity of benzaldehyde decreases. The maximum conversion was achieved at 24 h of reaction time.

The effect of temperature on the oxidation of benzyl alcohol was investigated by varying temperature in the range of 70–100 °C (Fig. 2.21). An optimum of 25.5 %

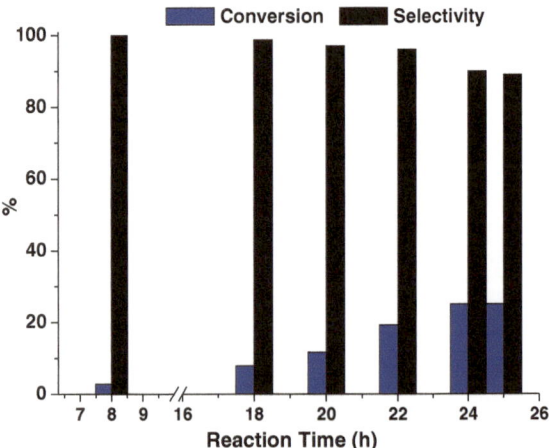

Fig. 2.20 % Conversion is based on benzyl alcohol; benzyl alcohol = 100 mmol, TBHP = 0.2 mmol, amount of catalyst = 25 mg, time = 24 h, temp = 90 °C

Fig. 2.21 % Conversion is based on benzyl alcohol; benzyl alcohol = 100 mmol, TBHP = 0.2 mmol, amount of catalyst = 25 mg, time = 24 h

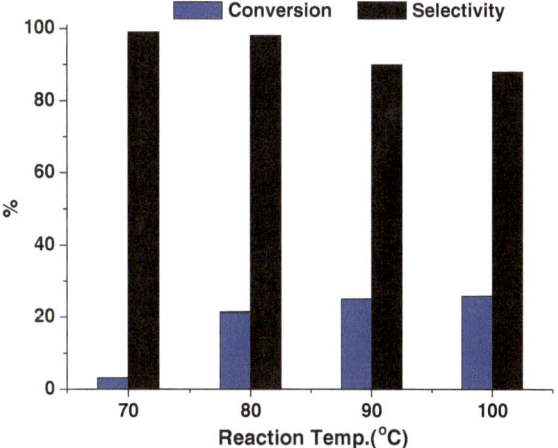

conversion was achieved at 90 °C temperature. When temperature was increased further to 100 °C there is no significant increase in conversion. This is due to over-oxidation of benzaldehyde to benzoic acid at elevated temperature. Hence, 90 °C was considered as optimum for the maximum conversion as well as selectivity.

The optimum conditions for maximum % conversion (25.5 %) of benzyl alcohol to benzaldehyde over 30 % PMo_{11}/Hβ are: loading of PMo_{11} = 30 %, amount of catalyst = 25 mg (0.227 wt%), time = 24 h, temperature = 90 °C. Under the same conditions the maximum % conversion over 30 % PMo_{11}/MCM-41 was found to be 28 %.

The control experiments with Hβ and PMo_{11} were also carried out under optimized conditions (Table 2.6). It can be seen from Table 2.6 that the supports are inactive towards the oxidation of benzyl alcohol indicating the catalytic activity is only due to PMo_{11}. The same reaction was carried out by taking the active amount of PMo_{11} (5.7 mg). It was found that the active catalyst gives 25.2 % conversion of benzyl alcohol with 91 % selectivity towards benzaldehyde. Similar obtained activity for supported catalysts indicate that PMo_{11} is the real active species. Thus, we are successful in supporting PMo_{11} onto MCM-41 as well as Hβ without any significant loss in activity and hence in overcoming the traditional problems of homogeneous catalysis.

Table 2.6 Control experiments over the catalyst and the support

Catalyst	Conversion %	Selectivity %
Hβ	<1	100
MCM-41	<1	100
30 % PMo_{11}/MCM-41	28.0	90
PMo_{11}	25.2	91
30 % PMo_{11}/Hβ	25.5	90

% conversion is based on benzyl alcohol; amount of PMo_{11} = 5.7 mg, amount of Hβ = 19.3 mg, time = 24 h, temp = 90 °C

Table 2.7 Oxidation of various alcohols over supported catalysts, under optimized conditions

Substrate	% Conversion	Products (% selectivity)	TON
[a]Benzyl alcohol	25.5	Benzaldehyde (90)	9,551
[a]Cyclohexanol	21	Cyclohexanone (>99)	8,580
[a]Cyclopentanol	20	Cyclopentanone (>99)	8,171
[a]1-Hexanol	9	1-hexanal (>99)	3,745
[a]1-Octanol	NC	1-Octanal (–)	–
[b]Benzyl alcohol	28	Benzaldehyde (90)	10,487
[b]Cyclohexanol	22	Cyclohexanone (>99)	8,989
[b]Cyclopentanol	20	Cyclopentanone (>99)	8,240
[b]1-Hexanol	15	1-hexanal (>99)	5,618
[b]1-Octanol	NC	1-Octanal (–)	–

% Conversion is based on benzyl alcohol; reactions of alcohols with O_2: TBHP = 0.2 mmol, amount of catalyst = 25 mg, temp = 90 °C, time = 24 h
[a] 30 % PMo_{11}/Hβ
[b] 30 % PMo_{11}/MCM-41

In order to explore the applicability of the method for a selective aerobic oxidation different alcohol substrates were studied (Table 2.7). Employing this system, benzylic, primary and secondary alcohols were oxidized to ketones or aldehydes at 90 °C with moderate to good conversions and excellent selectivity. It was observed from Table 2.7 that oxidation of secondary alcohol was easier as compared to primary alcohols. Also, it was observed that the long chain primary alcohol is very less reactive under the present reaction conditions. In all the cases we were able to achieve excellent TON for all the alcohols.

As described earlier, type of support does not play always merely a mechanical role but it can also modify the catalytic property of the catalysts. The difference in catalytic activity and product selectivity in oxidation reaction of benzyl alcohol may be due to the nature of supports such as structural mesoporosity and high specific surface area. Hence, the obtained results of 30 % PMo_{11}/Hβ were compared with 30 % PMo_{11}/MCM-41. From Table 2.7 it is seen that benzyl alcohol conversion is higher in the case of 30 % PMo_{11}/MCM-41. This might be due to difference in the surface area of the catalysts where surface area of 30 % PMo_{11}/MCM-41(485 $m^2 g^{-1}$) is higher compared to 30 % PMo_{11}/Hβ (362 $m^2 g^{-1}$). It is well known that the products distribution is affected by acidity of the catalyst. Generally, oxidation of benzyl alcohol gives benzaldehyde (major), with over oxidation product benzoic acid (minor). These types of over oxidation reactions are directly promoted by acidity of the catalyst. So, observed result i.e. lower selectivity (90 %) of benzal-dehyde for both the catalysts is attributed to acidity of the support, as it is well known that MCM-41 and zeolite Hβ both exhibit acidity. Hence the order of catalytic activity observed was 30 % PMo_{11}/MCM-41 > 30 % PMo_{11}/Hβ.

2.2.2.3 Heterogeneity Test

Heterogeneity test [12] was carried out for the oxidation of benzyl alcohol (Table 2.8). For the rigorous proof of heterogeneity, test was carried out by filtering catalyst from the reaction mixture at 90 °C after 18 h and the filtrate was allowed to react up to 24 h. The reaction mixture of 18 h and the filtrate were analysed by gas chromatography. No change in the % conversion as well as % selectivity was found indicating the present catalysts fall into category C i.e., active species does not leach and the observed catalysis is truly heterogeneous in nature. It also confirms that the reactions are not just auto-oxidation but the catalyst plays an important role for selective conversion of the substrates.

2.2.2.4 Catalytic Activity of Regenerated Catalysts

Oxidation of benzyl alcohol was carried out with the recycled supported catalysts, under the optimized conditions. The catalysts were removed from the reaction mixture after completion of the reaction by simple centrifugation; the first washing was given with dichloromethane to remove the products, then the subsequent washings were done by double distilled water and then dried at 100 °C, and the recovered catalyst was charged for the further run. No appreciable decrease in the conversion was observed up to two cycles (Table 2.9). The reused catalyst was further characterized by FT-IR and EDX analysis.

2.2.2.5 Characterization of Regenerated Catalysts

EDX elemental analysis performed on the recycled and fresh catalyst (Fresh: P = 0.37 %, Mo = 12.0 %; recycled: P = 0.33 %, Mo = 11.3 %) shows that the all elements in the recycled catalysts are retained. The FT-IR data for the fresh as well as the recycled catalysts are presented in Fig. 2.22. The FT-IR of recycled 30 % PMo_{11}/MCM-41 shows bands at 965 and 918 cm^{-1} assigned to P–O and Mo=O stretches, respectively. Similar observations can be made for recycled 30 % PMo_{11}/

Table 2.8 Heterogeneity test

Catalyst	Conversion %	Selectivity %
30 % PMo_{11}/Hβ (18 h)	8	98
Filtrate (24 h)	8	98
30 % PMo_{11}/MCM-41 (16 h)	12	98
Filtrate (24 h)	12	97

% Conversion is based on benzyl alcohol; amount of catalyst = 25 mg; temperature = 90 °C

Table 2.9 Recycling studies under the optimised conditions

Cycle[a,b]	Conversion (%)[a,b]	Selectivity (%)[a,b]	TON[a,b]
Fresh	28, 25.5	90, 90	10,487, 9,551
1	26, 25.0	89, 89	9,738, 9,364
2	25, 24.6	89, 87	9,363, 9,214

% Conversion is based on benzyl alcohol; Benzyl alcohol, O_2: 0.2 mmol TBHP, temp, 90 °C, time, 24 h, catalyst amount 25 mg
[a] 30 % PMo_{11}/MCM-41
[b] 30 % PMo_{11}/Hβ

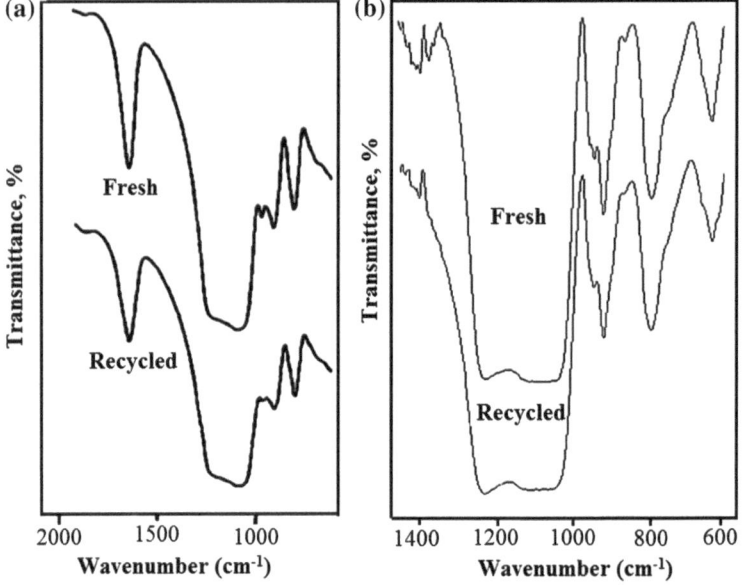

Fig. 2.22 FT-IR spectra of regenerated catalysts **a** 30 % PMo_{11}/MCM-41 and **b** 30 % PMo_{11}/Hβ

Hβ where almost all the bands are observed for PMo_{11}. No appreciable shift in the FT-IR band position of the regenerated catalyst compared to fresh 30 % PMo_{11}/Hβ indicates the retention of structure of PMo_{11} in the catalyst.

References

1. Pathan S, Patel A (2011) Dalton Trans 40:348–355
2. Patel A, Pathan S (2012) Ind Eng Chem Res 51(2):732–740
3. Richards R (ed) (2006) Surface and nanomolecular catalysis (Chap. 1). CRC Press, Taylor and Francis
4. Pathan S, Patel A (2014) Chem Eng J 243C:183–191

5. Vogel A (1951) A textbook of quantitative inorganic analysis, 2nd edn. Longmans, Green and Co., London

6. Okuhara T, Mizuno N, Misono M (1996) Adv Catal 41:113–252

7. Van Veen JAR, Sundmeijer O, Emeis CA, De Wit HJ (1986) J Chem Soc Dalton Trans 9:1825–1831

8. Edwards JC, Thiel CY, Benac B, Knifton JF (1998) Catal Lett 51:77–83

9. Shanmugam S, Viswanathan B, Varadarajan TK (2004) J Mol Catal A Chem 223:143–147

10. Balogh-Hergovicha E, Speier G (2005) J Mol Catal A Chem 230:79–83

11. Dhakshinamoorthy A, Alvaro M, Garcia H (2011) ACS Catal 1:48–53

12. Sheldon A, Walau M, Arends IWCE, Schuchurdt U (1998) Acc Chem Res 31:485–493

13. Cai Q, Luo ZS, Pang WQ, Fan YW, Chen XH, Cui FZ (2001) Chem Mater 13:258–263

14. Chidambaram M, Venkatesan C, Moreaub P, Finiels A, Ramaswamy AV, Singh AP (2002) Appl Catal A Gen 224:129–140

15. Narkhede N, Patel A, Singh S (2014) Dalton Trans 43:2512–2520

Chapter 3
Oxidation of Alcohols Catalyzed by Transition Metal Substituted Phosphomolybdates

Abstract Synthesis and characterization of mono transition metal-substituted Keggin-phosphomolybdates, $PMo_{11}M$ (M = Mn, Co, Ni, Cu) as well as their use as efficient catalysts for aerobic oxidation of alcohols in presence of TBHP as a radical initiator was explained. Optimization of reaction condition was also elucidated for oxidation benzyl alcohol as a model reaction. Moreover, oxidation of various alcohols such as cyclopentanol, 1-hexanol and 1-octanol with $PMo_{11}M$ was also presented under optimized conditions. All the catalysts showed good catalytic activity with excellent selectivity for the desired products as well as a higher TON. The catalytic activity of recycled catalysts was also described and it was found that catalysts are stable under present reaction conditions. The system not only catalyzes the reaction but also avoids the use of organic solvents as it was carried out under solvent free conditions. A probable reaction mechanism was also explained for the oxidation of alcohols.

Keywords Keggin phosphomolybdates · Transition metals · Oxidation · Alcohol · Mechanism

3.1 Experimental

All chemicals used were of A. R. grade. 12-Molybdophosphoric acid ($H_3PMo_{12}O_{40}$; PMo_{12}) sodium hydroxide, $Na_2MoO_4 \cdot 2H_2O$ $Co(CH_3COO)_2 \cdot 4H_2O$, $Mn(CH_3COO)_2 \cdot 4H_2O$, $Ni(CH_3COO)_2 \cdot 4H_2O$, CsCl, benzyl alcohol, cyclopentanol, cyclohexanol, 1-hexanol, 1-octanol, tertiary butylhydroperoxide (70 % aq. TBHP) and dichloromethane were obtained from Merck and used as received.

A. Patel and S. Pathan, *Polyoxomolybdates as Green Catalysts for Aerobic Oxidation*, SpringerBriefs in Green Chemistry for Sustainability, DOI 10.1007/978-3-319-12988-4_3

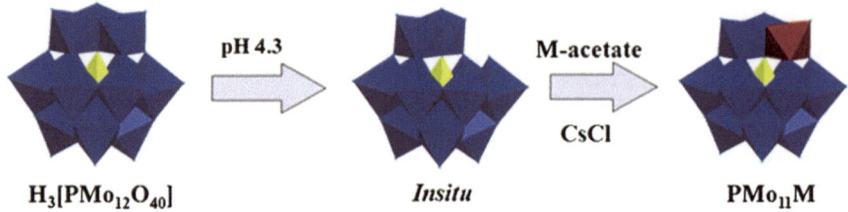

Scheme 3.1 Synthesis of PMo$_{11}$M (M = Co, Mn, Ni, Fe, Cu)

3.1.1 Synthesis of Transition Metal Substituted Phosphomolybdates (PMo$_{11}$M; M = Co, Mn, Ni, Cu) [1, 2]

H$_3$PMo$_{12}$O$_{40}$ (PMo$_{12}$, 1.825 g, 1 mmol) was dissolved in of water (10 mL) and the pH of the solution was adjusted to 4.3 using sodium hydroxide solution. Co(CH$_3$COO)$_2$·4H$_2$O (0.249 g, 1 mmol) dissolved in minimum amount of water was mixed with above hot solution. pH of solution was adjusted to 4.5 (pH 4.3 for nickel substituted phosphomolybdate). The solution was heated at 80 °C with stirring for 1 h and filtered hot. To the hot filtrate saturated solution of CsCl was added drop wise and allowed to stand. As obtained crystals were very poorly soluble in any solvent, no recrystallisation was carried out. The obtained dark reddish brown crystals were filtered, air dried and designated as PMo$_{11}$Co. Other transition metal (M = Mn, Ni and Cu) substituted phosphomolybdates were also synthesized according to the similar method, by taking corresponding metal-acetate salt. The obtained materials were designated as PMo$_{11}$Mn, PMo$_{11}$Ni, and PMo$_{11}$Cu respectively (Scheme 3.1).

Characterization specifications and catalytic reaction is similar as mentioned in Chap. 2 (Sects. 2.1.1.3 and 2.1.1.4).

3.2 Result and Discussion

3.2.1 Characterization of Transition Metal (Mn, Co, Ni, Cu)—Substituted Phosphomolybdates

PMo$_{11}$Co and PMo$_{11}$Mn were also characterized by single crystal XRD [1]. However, after several attempts, we are not succeeded in obtaining good quality crystals, suitable for single crystal X-ray analysis for PMo$_{11}$Ni and PMo$_{11}$Cu. Single crystal X-ray analysis of first raw transition metal substituted phosphomolybdates (PMo$_{11}$Co and PMo$_{11}$Mn,) showed structural disorders in Keggin unit. Thus, we are expecting the identical disordered structure for other PMo$_{11}$M

(M = Ni, Cu). However, all synthesized materials were characterized by other characterizations techniques also.

The complexes were isolated as the cesium salt after completion of the reaction and the remaining solution was filtered off. The filtrate was analyzed for molybdenum gravimetrically [3]. The observed proportion of Mo in the filtrate was 0.5 %, which corresponds to loss of one equivalent of Mo from $H_3PMo_{12}O_{40}$. The observed EDX values for the elemental analysis of the isolated compounds were in good agreement with the theoretical values.

For $PMo_{11}Co$: Anal Calc %: Cs, 25.88; Mo, 41.20; P, 1.20; Co, 2.30; O, 28.75.
Found %: Cs, 25.90; Mo, 41.29; P, 1.16; Co, 2.33; O, 28.69.
For $PMo_{11}Mn$: Anal Calc %: Cs, 26.00; Mo, 41.29; P, 1.21; Mn, 2.15; O, 28.79.
Found %: Cs, 26.34; Mo, 41.40; P, 1.18; Mn, 2.04; O, 28.61.
For $PMo_{11}Ni$: Anal Calc %: Cs, 26.03; Mo, 41.34; P, 1.21; Ni, 2.30; O, 28.84.
Found %: Cs, 26.12; Mo, 41.39; P, 1.24; Ni, 2.32; O, 28.89.
For $PMo_{11}Cu$: Anal Calc %: Cs, 25.91; Mo, 41.15; P, 1.21; Cu, 2.47; O, 28.70.
Found %: Cs, 26.11; Mo, 41.29; P, 1.19; Cu, 2.43; O, 28.96.

TGA of all of the catalysts (Table 3.1) show a weight loss of 4.2–4.83 % at 150 °C, corresponding to $7H_2O$ molecules. Similarly, DTA of all catalysts showed an endothermic peak around 130 °C due to water of crystallization. An exothermic peak in the region 415–430 °C indicates crystallization of the MoO_3 phase after decomposition of the Keggin unit.

Thus, based on the elemental analysis and thermal analysis, the formula of the complexes were proposed as $Cs_5[PM(H_2O)Mo_{11}O_{39}]6H_2O$ (M = Co, Mn, Ni, Cu).

The FT-IR data for PMo_{12}, $PMo_{11}Co$, $PMo_{11}Mn$ and $PMo_{11}Ni$ are shown in Table 3.2. The FT-IR of PMo_{12} show bands at 1,070, 965 cm^{-1} and 870 and 790 cm^{-1} corresponding to the symmetric stretching of P–O, Mo–O and Mo–O–Mo bonds, respectively. The FT-IR showed P–O bond frequency 1,050, 1,043, 1,046 and 1059 cm^{-1} for $PMo_{11}Co$, $PMo_{11}Mn$, $PMo_{11}Ni$ and $PMo_{11}Cu$ respectively. The shift in band position as compare to PMo_{12} indicates the introduction of transition metal into the octahedral lacuna. A shift in the stretching vibration of Mo = O and Mo–O–Mo for all three compounds was also observed indicating the complexation of the transition metals. An additional band at 480, 422, 442 and 508 cm^{-1} is attributed to the Co–O, Mn–O, Ni–O and Cu–O vibration, respectively. Thus, the FT-IR data clearly show the incorporation of transition metal into the Keggin framework.

Table 3.1 TG-DTA of $PMo_{11}Co$, $PMo_{11}Mn$ and $PMo_{11}Ni$

Catalyst	TGA (%Weight loss at 150 °C)	DTA	
		Endothermic	Exothermic
$PMo_{11}Co$	4.83	130	415
$PMo_{11}Mn$	4.21	125	430
$PMo_{11}Ni$	4.33	137	425
$PMo_{11}Cu$	4.31	130	410

Table 3.2 FT-IR spectra of PMo$_{12}$, PMo$_{11}$Co, PMo$_{11}$Mn, PMo$_{11}$Ni, PMo$_{11}$Cu

POMs	FT-IR frequencies cm^{-1}			
	P–O	Mo–O$_t$	Mo–O–Mo	M–O
PMo$_{12}$	1,070	965	870, 790	–
PMo$_{11}$Co	1,050	944	866, 797	480
PMo$_{11}$Mn	1,043	950	873, 800	422
PMo$_{11}$Ni	1,046	949	872, 799	442
PMo$_{11}$Cu	1,059	961	883, 801	508

Further, the presence of Co(II), and Mn(II) in the synthesized compounds was confirmed by ESR. The full range (3,200–2,000 G) X-band Room temperature ESR spectra (Fig. 3.1) for PMo$_{11}$Co, and PMo$_{11}$Mn was recorded. ESR spectra of PMo$_{11}$Co shows eight hyperfine signals (Co^{2+}; I = 7/2), confirming the presence of paramagnetic Co(II). The observed g value of ∼2.66 shows that Co(II) is in octahedral or distorted octahedral environment.

ESR spectra of PMo$_{11}$Mn shows six signals (Mn^{2+}; s = 5/2), indicates the presence of Mn(II) with octahedral or distorted octahedral symmetry. It is reported that the g value of ∼2 attributed to octahedral or distorted octahedral environment of Mn(II) [4]. The obtained g value (2.03) is in good agreement with a reported one and confirms the presence of paramagnetic Mn(II). Similarly, ESR spectra of PMo$_{11}$Cu (Cu^{2+}; S = 1/2, I = 3/2) shows typical lines pattern for distorted octahedral geometry with g$^{\parallel}$ = 2.40 and g$^{\perp}$ = 2.10.

Fig. 3.1 ESR spectra of **a** PMo$_{11}$Co and **b** PMo$_{11}$Mn, **c** PMo$_{11}$Cu

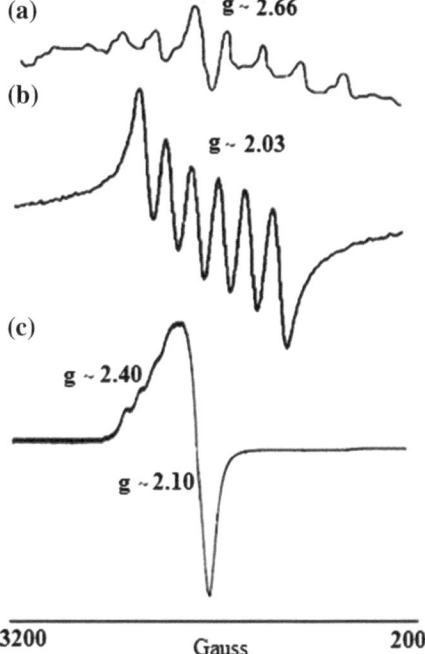

(a)   g ∼ 2.66

(b)   g ∼ 2.03

(c)   g ∼ 2.40

         g ∼ 2.10

3200   Gauss   2000

3.2.2 Catalytic Activity

In order to optimize the conditions, a detail study was carried out on oxidation of benzyl alcohol over $PMo_{11}Co$. To ensure the catalytic activity, all reactions were carried out without catalyst and no oxidation takes place. The reaction was carried out with 20 mg of catalysts for 24 h at 90 °C. Generally, benzyl alcohol on oxidation gives benzaldehyde and benzoic acid. However, benzaldehyde was obtained as the major oxidation product in the present case.

In order to determine the optimum temperature the reaction was investigated at three different temperatures 70, 90 and 110 °C, keeping other parameters fixed (20 mg catalyst amount, reaction time 24 h). The result for the same is presented in Fig. 3.2.

The results show that conversion increased with increasing temperature from 70 to 110 °C. At the same time, on increasing temperature from 90 to 110 °C, drastic decrease in selectivity of benzaldehyde was observed. This is due to over oxidation of benzaldehyde to benzoic acid at elevated temperature. So the temperature of 90 °C was found optimized for further studies.

The effect of amount of the catalyst on the conversion was studied and the obtained result is shown in Fig. 3.3.

With increases in the amount of catalysts i.e. concentration of metal contains, % conversion also increases. This suggests that transition metal functions as active sites for oxidation. It is very interesting to observe the difference in the selectivity of the products with increase in the concentration of the catalyst. It is observed from Fig. 3.3 that with lower amount of catalysts, >96 % selectivity of benzaldehyde is obtained. On increasing amount of catalyst (more than 20 mg), selectivity for benzaldehyde decreases. This may be due the fact that with increase in the amount of the active species the reaction becomes fast which favours the conversion of the

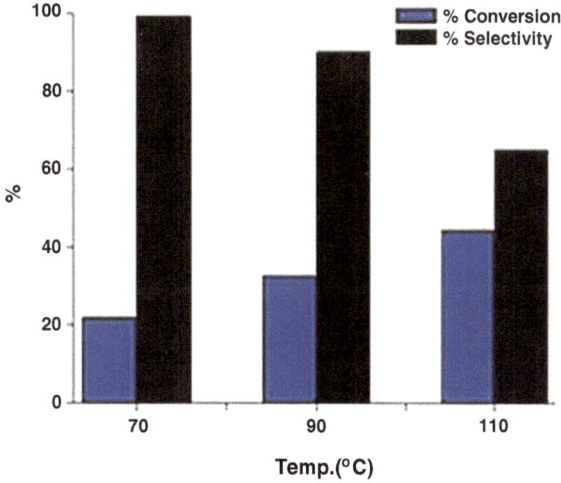

Fig. 3.2 % Conversion is based on benzyl alcohol; benzyl alcohol = 100 mmol, TBHP = 0.2 %, time = 24 h, amount of $PMo_{11}Co$ = 20 mg

Fig. 3.3 % Conversion is based on benzyl alcohol; benzyl alcohol = 100 mmol, TBHP = 0.2 %, time = 24 h, temp = 90 °C

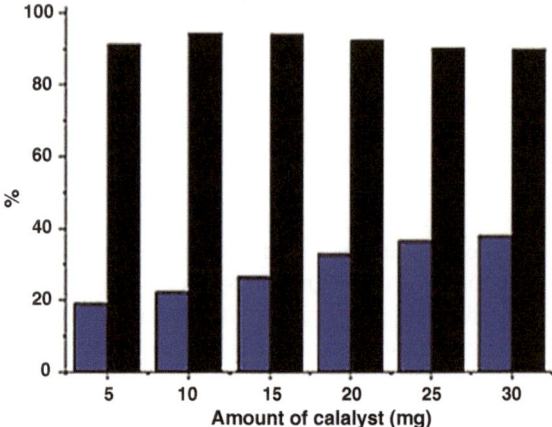

formed benzaldehyde to benzoic acid. Thus, amount of catalyst was optimized to 20 mg for optimum conversion and selectivity.

The percentage of conversion was monitored at different reaction times and result is presented in Fig. 3.4.

It is seen from Fig. 3.4, that with increase in reaction time, the % conversion also increases. Initial conversion of benzyl alcohol increased with the reaction time. This is due to the reason that more time is required for the formation of reactive intermediate (substrate + catalyst) which is finally converted into the products. It is seen that 32.6 % conversion of benzyl alcohol with 92.3 % selectivity of benzaldehyde was observed at 24 h. When the reaction was allowed to continue after 24 h, no significant change in conversion was observed, but selectivity of benzaldehyde was decreased. This is because of over oxidation of benzaldehyde to benzoic acid. So, reaction time was optimized as 24 h.

Fig. 3.4 % Conversion is based on benzyl alcohol; benzyl alcohol = 100 mmol, TBHP = 0.2 %, amount of $PMo_{11}Co$ = 20 mg, temp = 90 °C

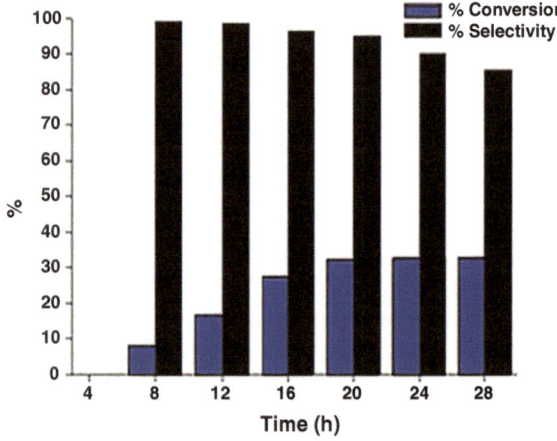

The optimum condition for optimum conversion of benzyl alcohol and selectivity of benzaldehyde over $PMo_{11}Co$ is; 20 mg of catalyst, 24 h reaction time, reaction temperature 90 °C.

In order to see the effect of substitution of metal in PMo_{11} oxidation of benzyl alcohols was also carried out using $PMo_{11}M$ (M = Mn, Co, Ni, Cu) under optimized conditions and results are presented in Table 3.3.

It is seen from Table 3.1 that, the order of activity of catalysts for oxidation of benzyl alcohol is $PMo_{11}Co \approx PMo_{11}Ni > PMo_{11}Mn > PMo_{11}Cu$. It is seen from the data that $PMo_{11}Cu$ is least active among $PMo_{11}M$ for benzyl alcohol oxidation under present reaction condition. Thus, oxidation of other alcohols was not carried for the same.

As seen from Table 3.3 that, as benzyl alcohol oxidation, similar trend in activity of $PMo_{11}M$ (M = Co, Ni, Mn) was observed for oxidation of other alcohols such as cyclopentanol, 1-hexanol and 1-octanol i.e. $PMo_{11}Co$ shows better activity as compare to $PMo_{11}Mn$ and $PMo_{11}Ni$. It was observed from Table 3.3 that, oxidation of secondary alcohol is easier as compare to primary alcohols. The observed trend is in good agreement with reported in art [5]. It is also observed that, in all cases, very good selectivity of desired product with high TON is obtained. At the same time, present catalytic system is also not applicable to less reactive long chain primary alcohol such as 1-octanol. For the present catalytic system, the reactivity of the alcohols was found in the order primary < cyclic secondary < aromatic.

Table 3.3 Oxidation of alcohols with O_2 using $PMo_{11}M$

Catalyst	Alcohols	% Conversion	Products (%selectivity)	TON
$PMo_{11}Co$	Benzyl alcohol	32.6	Benzaldehyde (92.3)	4,281
	Cyclopentanol	29.8	Cyclopentanone (100)	3,913
	Cyclohexanol	29.2	Cyclohexanone (100)	3,834
	1-Hexanol	20.6	1-Hexanal (100)	2,705
	1-Octanol	NC	–	–
$PMo_{11}Mn$	Benzyl alcohol	28.3	Benzaldehyde (93.6)	3,628
	Cyclopentanol	24.3	Cyclopentanone (100)	3,115
	Cyclohexanol	23.7	Cyclohexanone (100)	3,038
	1-Hexanol	15.9	1-Hexanal (100)	2,038
	1-Octanol	NC	–	–
$PMo_{11}Ni$	Benzyl alcohol	31.9	Benzaldehyde (92.3)	4,075
	Cyclopentanol	27.6	Cyclopentanone (100)	3,526
	Cyclohexanol	24.1	Cyclohexanone (100)	3,078
	1-Hexanol	17.7	1-Hexanal (100)	2,261
	1-Octanol	NC	–	–
$PMo_{11}Cu$	Benzyl alcohol	17.7	Benzaldehyde (99.4)	1815

% Conversion is based on substrate; substrate = 100 mmol, TBHP = 0.2 %, amount of catalysts = 20, 50 mg, temperature = 90 °C, NC = no conversion

Table 3.4 Controlled experiment for oxidation of benzyl alcohol with CsPMo$_{11}$, CsPMo$_{12}$, PMo$_{11}$M under optimized conditions

Entry	Catalyst	% Conversion	% Selectivity	
			Benzaldehyde	Other
1	CsPMo$_{11}$	13.4	91.2	8.8
2	CsPMo$_{12}$	6.2	87.2	12.8
3	PMo$_{11}$M	17–33	92–99	1–8

% Conversion is based on benzyl alcohol, benzyl alcohol = 100 mmol, O$_2$, TBHP = 0.2 %, temp = 90 °C, time = 24 h, amount of catalyst = 25 mg (PMo$_{11}$M), 18 mg (CsPMo$_{12}$), 27 mg PMo$_{11}$

A controlled experiment for oxidation of benzyl alcohol with CsPMo$_{11}$ and CsPMo$_{12}$, and PMo$_{11}$M (M = Co, Mn, Ni) was carried out under optimized conditions and results are presented in Table 3.4.

It is seen from Table 3.4 that in CsPMo$_{11}$ and CsPMo$_{12}$ catalyzed oxidation reactions, very low conversion of substrates and with 80–90 % selectivity of benzaldehyde was obtained whereas, PMo$_{11}$M, show good conversion i.e. >17 % more conversion as well as excellent selectivity for benzaldehyde was observed. Controlled experiment shows that PMo$_{11}$M (M = Co, Mn, Ni) are better catalysts as compare to their parent (CsPMo$_{12}$) and lacunary counter-part (CsPMo$_{11}$). From experiment it will be also concluded that although transition metal can act as active centre for oxidation, possibility of involvement of the Mo species cannot be ruled out.

3.2.2.1 Heterogeneity Test

Heterogeneity test was carried out for the oxidation of benzyl alcohol (Fig. 3.5) over PMo$_{11}$Co as examples. For the rigorous proof of heterogeneity, a test [6] was carried out by filtering catalyst from the reaction mixture at 90 °C after 12 h and the filtrate was allowed to react up to 28 h.

Fig. 3.5 % Conversion is based on substrate benzyl alcohol; amount of PMo$_{11}$Co = 20 mg; substrate 100 mmol, O$_2$, TBHP = 0.2 %, temperature 90 °C

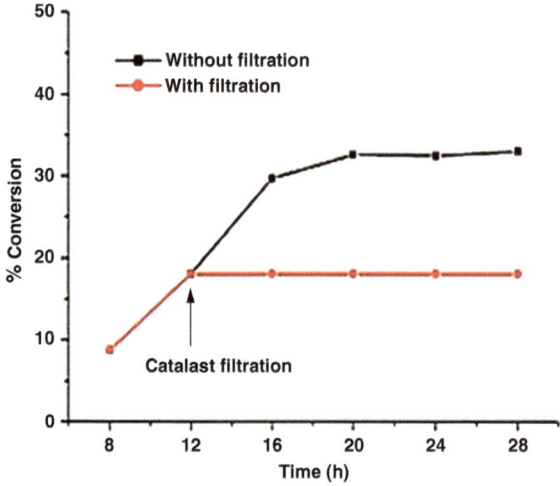

The reaction mixture of 12 h and the filtrate were analyzed by gas chromatography. Similar test was carried for $PMo_{11}M$ (M = Mn, Ni, Cu). No change in the % conversion as well as % selectivity was found indicating the present catalysts fall into category C [6] i.e., active species does not leach and the observed catalysis is truly heterogeneous in nature.

3.2.2.2 Catalytic Activity of Regenerated Catalysts

The catalyst was recycled in order to test for its activity as well as its stability. The catalysts remain insoluble under the present reaction conditions. The decomposition or leaching of metal contain from $PMo_{11}M$ was confirmed by carrying out an analysis of the used catalyst (EDX) as well as the product mixtures (AAS). For all catalysts, the analysis of the used catalyst did not show appreciable loss in the metal content as compared to the fresh catalyst. Analysis of the product mixtures shows that if any metal was present it was below the detection limit, which corresponded to less than 1 ppm. These observations strongly suggest that the present catalyst is truly heterogeneous in nature.

$PMo_{11}Co$ was separated easily by simple filtration followed by washing with dichloromethane and dried at 100 °C. Oxidation reaction was carried out with the regenerated catalysts, under the optimized conditions. The data for the catalytic activity is represented in Table 3.5. It is seen from the Table 3.5 that there was no significant change in conversion as well as selectivity. This shows that the catalysts are stable under present reaction conditions. Similarly, other $PMo_{11}M$ can be regenerated and reused successfully up to 4 cycles.

3.2.2.3 Characterization of Regenerated Catalysts

The separated catalysts were washed with dichloromethane and then water and dried at 100 °C. The obtained recycled catalysts were characterized for FT-IR. The FT-IR spectra for the fresh as well as the regenerated catalysts are represented in Fig. 3.6.

Table 3.5 Oxidation of benzyl alcohol with fresh and regenerated catalysts

Cycle	Conversion (%)	Selectivity (%)
		Benzaldehyde
Fresh	32.6	92.3
1	32.6	92.0
2	32.6	92.2
3	32.4	92.2
4	32.3	92.1

% Conversion is based on substrate; substrate = 100 mmol, TBHP = 0.2 %, temp = 90 °C, time = 24 h amount of $PMo_{11}Co$ = 20 mg

Fig. 3.6 FT-IR spectra of
a fresh and **b** recycled
PMo$_{11}$Co

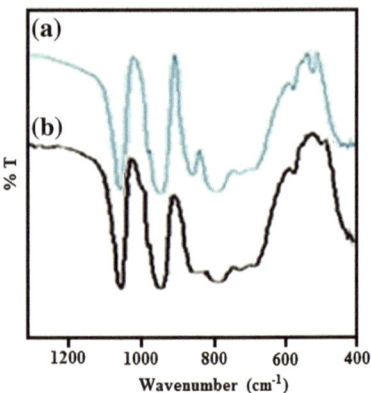

No appreciable shift in the FT-IR band position of regenerated PMo$_{11}$Co compare to fresh, indicates the retention of Keggin-type structure i.e. PMo$_{11}$M are stable under present reaction condition and can be re-used up to four cycles.

3.2.2.4 Probable Reaction Mechanism

In order to study the reaction mechanism the same sets of reactions were carried under two different conditions; (i) benzyl alcohol + oxidant + TBHP and (ii) benzyl alcohol + oxidant + PMo$_{11}$M. In both the cases the reaction did not progress significantly. These observations indicate that the liberation of O$_2$ from TBHP was not sufficient to proceed the reaction as well as activation of M^{2+} to M^{3+} being necessary for provoking the reaction under the optimized conditions. Hence it may be concluded that in present study TBHP acts as a radical initiator only.

For these types of reaction, inhibition experiment was carried out and radical mechanism was already established [7]. In order to confirm radical mechanism for present catalytic system, similar experiment was carried out by the use of excess i-PrOH as radical terminator. Based on the results, a tentative reaction mechanism for oxidation of benzyl alcohol with O$_2$ in presence of TBHP as an initiator is proposed in Scheme 3.2.

$$PMo_{11}M^n + \textit{tert}\text{-BuOOH} \rightarrow PMo_{11}M^{n+1} + \textit{tert}\text{-BuO}^{\cdot} + OH^-$$

It has been reported that, in case of transition metal substituted polyoxotung-states, transition metals behaved as active centre for catalysis [8, 9]. It has been also reported that for TMSPOMs catalysts containing metal cations in low valency states and involving O$_2$ as an oxidant always follow the radical chain mechanism induced by M–O$_2$ intermediate [10, 11].

In the present catalytic system, the mechanism is expected to follow the same path. It is expected that, reaction of M^{2+} with TBHP cause oxidization of M^{2+} to M^{3+}

$$PMo_{11}M^n + \textit{tert}\text{-BuOOH} \longrightarrow PMo_{11}M^{n+1} + \textit{tert}\text{-BuO}\cdot + OH^-$$

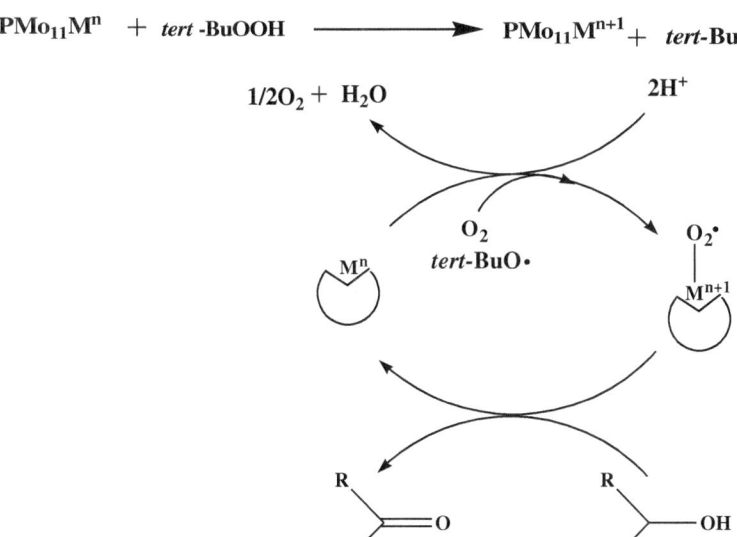

Scheme 3.2 Proposed reaction mechanism for oxidation of alcohol using O_2

in situ. The activation of this specie takes place with radical $(\textit{tert} - BuO^{\cdot})$ generated during decomposition of TBHP and attack of O_2 simultaneously, which results in formation of activated specie $^{\cdot}OOM^{3+}PMo_{11}$. This active specie $OOM^{3+}PMo_{11}$ is responsible for oxidation of alcohols to corresponding carbonyl compounds (Scheme 3.2). But at the same time, as describe in the controlled experiment, addenda atoms (Mo species) may also be involved the catalytic cycle.

References

1. Patel A, Pathan S (2012) J Coord Chem 65:3122–3132
2. Pathan S, Patel A (2013) Appl Catal A Gen 459:59–64
3. Vogel A (1951) A textbook of quantitative inorganic analysis, 2nd edn. Longmans, Green and Co., London
4. Nowinska K, Waclaw A, Masierak W, Gutsze A (2004) Catal Lett 92:157–162
5. Balogh-Hergovicha E, Speier G (2005) J Mol Catal A Chem 230:79–83
6. Sheldon A, Walau M, Arends IWCE, Schuchurdt U (1998) Acc Chem Res 31:485–493
7. Pathan S, Patel A (2014) Catal Sci Technol 4:648–656
8. Hill CL, Brown RB Jr (1986) J Am Chem Soc 108:536–538
9. Wang J, Yan L, Li G, Wang X, Dinga Y, Suoa J (2005) Tetrahedron Lett 46:7023–7027
10. Xinrong L, Jinyu X, Huizhang L, Bin Y, Songlin J, Gaoyang X (2000) J Mol Catal A Chem 161:163–169
11. Zhang X, Chen Q, Duncan DC, Lachicotte RJ, Hill CL (1997) Inorg Chem 36:4381–4386

Chapter 4
Conclusive Remarks

In the present book, we have discussed an overview of a part of research conducted in our group on the development of heterogeneous catalysts for oxidation reactions under solvent free green reaction conditions. Unique properties of POMs permit designing at molecular level. Special attention was focused on the designing and application of POMs based on phosphomolybdates. As a part of development of new phosphomolybdates materials (1) synthesis and isolation of PMo_{11} was described. Also, stabilization of PMo_{11} was successfully done by supporting it to suitable supports (2) one-pot synthesis of $PMo_{11}M$ (M = Mn, Co, Ni, Cu) was also discussed. These procedures open up new synthetic routes for development of other POMs based material.

The synthesized materials have been proved to be successful heterogeneous catalysts for oxidation alcohol with molecular oxygen under solvent free green conditions. In all cases, moderate-to-good conversion with very good selectivity of desire products as well as high TON was achieved. All the catalysts can be successfully regenerated and reused up to 4 cycles without any significant loss in conversion as well selectivity. The advantages of reusable and sustainable catalysts with molecular oxygen for oxidation under solvent free conditions make this methodology interesting from an economic and an ecological point of view. However, present catalytic systems are not applicable for oxidation of long chain (C8 onwards) alcohols. Thus, future target will need stagiest to overcome such problems so as to open up a new possibility for the use of these catalysts for many practical oxidations.

© The Author(s) 2015

A. Patel and S. Pathan, *Polyoxomolybdates as Green Catalysts for Aerobic Oxidation*,
SpringerBriefs in Green Chemistry for Sustainability,
DOI 10.1007/978-3-319-12988-4_4